Springer Series on Polymer and Composite Materials

The "Springer Series on Polymer and Composite Materials" publishes monographs and edited works in the areas of Polymer Science and Composite Materials. These compound classes form the basis for the development of many new materials for various applications. The series covers biomaterials, nanomaterials, polymeric nanofibers, and electrospun materials, polymer hybrids, composite materials from macro- to nano-scale, and many more; from fundamentals, over the synthesis and development of the new materials, to their applications. The authored or edited books in this series address researchers and professionals, academic and industrial chemists involved in the areas of Polymer Science and the development of new Materials. They cover aspects such as the chemistry, physics, characterization, and material science of Polymers, and Polymer and Composite Materials. The books in this series can serve a growing demand for concise and comprehensive treatments of specific topics in this rapidly growing field. The series will be interesting for researchers working in this field and cover the latest advances in polymers and composite materials. Potential topics include, but are not limited to:

Fibers and Polymers:

- Lignocellulosic biomass and natural fibers
- Polymer nanofibers
- Polysaccharides and their derivatives
- Conducting polymers
- Surface functionalization of polymers
- Bio-inspired and stimuli-responsive polymers
- Shape-memory and self-healing polymers
- Hydrogels
- Rubber
- Polymeric foams
- Biodegradation and recycling of polymers

Bio- and Nano- Composites:

- Fiber-reinforced composites including both long and short fibers
- Wood-based composites
- Polymer blends
- Hybrid materials (organic-inorganic)
- Nanocomposite hydrogels
- Mechanical behavior of composites
- The Interface and Interphase in polymer composites
- Biodegradation and recycling of polymer composites
- Applications of composite materials

Ayyappa Atmakuri · Arvydas Palevicius · Giedrius Janusas

Fiber-Based Hybrid Composites for Engineering

Springer

Ayyappa Atmakuri
Faculty of Mechanical
Engineering and Design
Kaunas University of Technology
Kaunas, Lithuania

Arvydas Palevicius
Faculty of Mechanical
Engineering and Design
Kaunas University of Technology
Kaunas, Lithuania

Giedrius Janusas
Faculty of Mechanical
Engineering and Design
Kaunas University of Technology
Kaunas, Lithuania

ISSN 2364-1878 ISSN 2364-1886 (electronic)
Springer Series on Polymer and Composite Materials
ISBN 978-3-031-93143-7 ISBN 978-3-031-93144-4 (eBook)
https://doi.org/10.1007/978-3-031-93144-4

This Springer imprint is published by the registered company Springer Nature Switzerland AG
The registered company address is: Gewerbestrasse 11, 6330 Cham, Switzerland

Preface

The field of composite materials has undergone marked advancements in recent years, driven by the pursuit of sustainable, high-performance alternatives to traditional materials. Within this dynamic context, natural fiber-based hybrid composites have surfaced as a viable solution, presenting an exceptional balance of mechanical strength, environmental sustainability, and cost-efficiency. This monograph brings together a collection of scientific research papers, offering an in-depth exploration of the latest developments, methodologies, and future possibilities of natural fiber composites in the dynamic field of materials science and engineering. Within these pages, it discovers the intricate world of fabrication techniques for hybrid composites based on natural fibers and various matrix materials. This monograph unravels the secrets behind the enhanced mechanical properties of natural fiber hybrids, showcasing remarkable tensile strength, flexural strength, and modulus improvements. It explores the impact of alkaline treatments on the mechanical properties of reinforcement materials through detailed morphological studies and FTIR analyses. Comparative studies between epoxy-matrix-based and EcoPoxy (Bio resin) matrix-based hybrid composites reveal intriguing insights, while investigations into basalt and E-glass woven-fiber-based epoxy hybrid composites with graphite filler material showcase superior mechanical properties. Moreover, this monograph goes beyond presenting findings; it sets the stage for future research. Proposals for further investigations include exploring diverse chemical treatments for natural fibers, studying the effects of various filler materials, analyzing moisture absorption behavior, and expanding the application scope of natural fiber hybrid composites. Whether you're an experienced researcher, an engineering enthusiast, or a student entering the captivating realm of mechanical engineering or materials science, this monograph will be a valuable resource. This monograph takes a profound journey into the advancements, challenges, and potential applications of natural fiber hybrid composites, shaping the future of sustainable engineering solutions and innovative materials with advancements in natural fiber hybrid composites. This monograph can also be used as a methodological tool for lectures in advanced courses on hybrid composites.

Kaunas, Lithuania
2025

Ayyappa Atmakuri
Arvydas Palevicius
Giedrius Janusas

Contents

Chapter 1
Introduction

Abstract Natural fiber-reinforced polymer composites (NFCs) are gaining prominence as eco-efficient alternatives to conventional synthetic composites, driven by their biodegradability, low density, renewability, and competitive mechanical properties. Among recent developments, hybrid natural fiber composites comprising two or more natural or synthetic fibers embedded within a polymer matrix have demonstrated enhanced performance by leveraging the complementary properties of different fibers. These composites address key limitations of single-fiber systems, such as poor moisture resistance, dimensional instability, and low interfacial adhesion with hydrophobic matrices. The mechanical behavior of hybrid composites is largely determined by factors such as fiber type, length, orientation, surface morphology, and the quality of the fiber–matrix interface. To improve interfacial bonding, various surface modification techniques—such as alkali treatment, plasma activation, and chemical grafting—are employed to increase surface roughness, remove hemicellulose and waxes, and introduce functional groups that enhance interfacial shear strength (IFSS). These treatments significantly improve stress transfer, leading to composites with superior tensile, flexural, and impact properties, along with improved thermal and environmental durability. Hybrid natural fiber composites have found applications in diverse sectors including automotive (interior panels, bumpers), construction (insulative boards, paneling), biomedical (scaffolds, drug delivery systems), and emerging microfluidic devices, where lightweight, strength-to-weight efficiency and environmental safety are critical. With continued innovation in hybridization strategies, interfacial engineering, and processing techniques, Natural fiber hybrid composites present a robust path forward for replacing synthetic fiber composites in structural and functional applications aligned with sustainable development goals.

In recent years, significant interest has emerged in developing composites utilizing natural fibers enhanced by a polymer matrix. Natural fiber hybrid composites (NFCs) are a type of material in which a polymer resin matrix fortifies natural fibers [1]. The materials gain attention for sequestering carbon dioxide, mitigating pollution, and not releasing harmful gases [2], while natural fibers provide benefits over synthetic ones,

A. Atmakuri et al., *Fiber-Based Hybrid Composites for Engineering*, Springer Series on Polymer and Composite Materials, https://doi.org/10.1007/978-3-031-93144-4_1

such as biocompatibility, affordability, high strength, and minimal environmental impact [3–5]. The global natural fiber market is projected to grow at an annual compound growth rate of 9.4% between 2017 and 2022, and the growth rate of synthetic fibers, such as those made from banana, pineapple, sisal, hemp, flax, kenaf, and jute, is regarded as a feasible alternative to conventional synthetic fibers [6].

Natural fiber composites are utilized across numerous fields such as construction, biomedical sciences, automotive, microfluidics, energy, shipping, and military. Within the microfluidics industry, for instance, these composites are employed to create lightweight components like cell culture systems, drug delivery platforms, and hospital diagnostics [7]. Despite their complexity, they are not currently a primary research focus. In general, composite materials consist of two or more constituents, such as a matrix and reinforcement. The reinforcement encased in a matrix, which serves as a binder [8], is a substance that adds extra strength to the composite [8–10]. Composites are significant due to their enhanced physical, chemical, and mechanical properties compared to individual components. They can be grouped into polymer, ceramic, and metal matrix composites based on their matrix [11]. The effectiveness of natural fiber composites primarily hinges on aspects like length, configuration, shape, origin, and the interaction at the fiber-matrix interface. Fibers may be termed continuous or discontinuous based on their length-to-diameter ratio. Reinforcements, categorized into woven or non-woven arrangements, are available in various forms, such as laminates, fibrous structures, and particulates, and are divided into natural and synthetic types based on material origin and availability [12–14]. Recent studies have demonstrated that the evolution of hybridization significantly affects the characteristics of composite materials [15]. These studies explored innovative techniques for merging various materials, such as combining two distinct types of natural fibers, integrating them with synthetic fibers, or incorporating two types of synthetic fibers into the same matrix material, to enhance overall quality and ensure improvements in mechanical properties over monofilament composites. By utilizing different sizes, filler types, and fiber orientations, a variety of hybrid composites can be developed. Hybridization using various types of material fibers can address the shortcomings of individual components [16]. The hybrid effect in composite materials depends on the mechanical and chemical attributes of the fibers, so its effectiveness is diminished compared to composite materials made of single fibers [17]. In recent decades, extensive studies have been undertaken to investigate the combination of synthetic and plant fibers in composite materials [18, 19].

Nonetheless, hybrid composites strengthened by natural fibers have recently been the subject of less research, as most attention has been given to combinations of natural and synthetic fibers. An important example of such combinations includes the use of carbon, glass, and aramid mixed with hemp, flax, bananas, and pineapples [20, 21]. Hybrid fiber reinforcement composites are seen as superior to single fiber composites. Their enhanced mechanical properties and cost-effectiveness offer additional applications [22]. Consequently, hybrid composites have replaced artificial fiber composites due to their outstanding characteristics, especially in load-bearing and construction applications [23, 24]. The use of natural fiber composites in various industries such as household appliances, sports gear, furniture, packaging, and lightweight goods [25, 26], is on the rise. Typically, natural fiber composites are short fiber composites characterized by irregular fiber lengths and uneven reinforcement orientation distribution [27]. The elasticity of discontinuous fiber composites is similar to that of continuous fiber reinforced composites, with only minor differences in elastic moduli and resistances [28]. However, the flexibility in choosing fabrication techniques is greater. Discontinuous fiber composites can be manufactured using reverse plastic processing techniques, allowing for potentially easier mass production [29].

In the realm of composite materials, both reinforcements and the matrix preserve their distinct physical and chemical characteristics. Nevertheless, when combined, these components generate singular properties that neither can achieve on their own [30]. The primary aim of these materials is to find a balance of properties by merging different components. Additional elements like binding agents, coatings, and fillers can be included in composites. Agents, coatings, and nanoparticle fillers are applied to fibers to enhance compatibility with the matrix, improve wetting and bonding at the reinforcement-matrix interface, and enhance the mechanical and physical attributes of composites [31, 32]. Although natural fiber components have superior mechanical properties, they also possess certain drawbacks like insufficient adhesion, ductility, moisture resistance, and internal stress [33–36]. Such deficiencies might undermine the structural integrity and mechanical performance of composite materials. To address these issues, various efforts have been made using hybrid polymer composites reinforced with natural fibers. Alkaline and heat treatments are applied to these materials, and diverse manufacturing techniques are developed to thoroughly understand production errors. Hybrid composites with varying weight percentages have been produced to improve mechanical properties and suggest applications in sustainable technologies like microfluids. These applications encompass disposable devices, components of microfluid systems (e.g., substrates), and device packaging. In recent years, there has been significant interest in utilizing natural fibers, both renewable and non-renewable—including banana, orange, pineapple, hemp, sunflower, flax, bamboo, and jute—to create composite materials [37]. Plants that yield cellulose fibers can be categorized into several types, such as bast fibers (flax, hemp, jute, kenaf), seed fibers (kapok, coir, cotton), leaf fibers (abaca, pineapples, and sips), grass fibers, and bark fibers (wheat, rice, corn), kernel fibers (hemp, jute, kenaf), and diverse sources like roots and wood [38, 39]. Natural fibers provide numerous advantages over synthetic fibers, including renewability, biodegradability, cost efficiency, minimal damage during processing, lightweight, improved surface finishes for composite molding components, ease of processing, favorable mechanical properties such as elasticity and flexible moduli, and fewer health risks [40]. In many ways, the advantages of natural fibers surpass those of synthetic fibers [41].

In the 1960s, Toyota made notable strides in incorporating fenders and roof lids crafted from soy protein eco-plastic into its vehicles [42]. Historically considered minor compared to synthetic composites, research and development of composite materials from renewable and sustainable sources are now crucial for future industrial practices due to the heightened emphasis on sustainable growth [43]. The increasing societal demand for eco-friendly and biodegradable materials has significantly driven progress in the science of composite materials. The biodegradability of natural fibers plays a significant role in the application of polymer materials [44]. By designing lightweight and strong biomaterials, advanced lightweight composites can be engineered as alternatives to traditional synthetic structures or metal components for sustainable purposes [45]. Natural fibers offer benefits, but their use in composites for structural applications is challenging due to drawbacks such as high susceptibility to moisture absorption during manufacturing, thermal and mechanical degradation, incompatibility with some polymer matrices, and low resistance to microbial degradation [46]. To address these issues, innovations in hybrid composite materials incorporating natural fibers may offer solutions. By introducing varied types of fibers, the promising approach of hybrid composite materials, which combines natural and synthetic fibers, addresses the limitations associated with composites reinforced by single fiber types [47, 48]. Interfacial adhesion between reinforcement and matrix materials is critical to the mechanical performance of composite materials [49]. Insufficient adhesion between natural fibers and matrices, due to their distinctive chemical composition, water resistance, and surface characteristics, remains a significant barrier to the industrial application of natural fibers [50]. The poor bond between the natural fiber and the matrix can permit water ingress, revealing the hydrophilic fiber group to water molecules. Consequently, this scenario may lead to the deterioration of the composite and reduced mechanical strength [51, 52].

The optimal mechanical functioning of composite materials relies on enhancing the stress transfer ability of the fiber-matrix interface, which is closely related to the level of adhesion. Several factors contribute to interface adhesion, including interaction force, chemical bonding, and residual stress [53]. In hybrid composites, there are two types of interfaces, particularly interfacing interfaces and interfacing interfaces. The quality of adhesion in both interfaces significantly influences the mechanical properties of the hybrid composite. However, interaction interfaces are distinctly important in hybrid composites as they impact the overall performance of materials [54]. Recent studies have indicated that altering the surfaces of fibers can improve the adhesion characteristics of composite materials. Chemical treatments involving alkaline or alkaline solutions are among the most effective surface modification techniques, significantly affecting fiber morphology [55]. A major hurdle in the industrial use of natural fibers is their incompatibility with polymer matrices. Natural fibers vary in chemical makeup and surface properties, hindering the formation of strong interface adhesion to hydrophobic polymer matrices, leading to inferior mechanical composites [56]. Natural fibers primarily consist of cellulose, hemicellulose, wax, lignin, and pectin, and their relative proportions can vary depending on growth conditions [57]. Cellulose, hemicellulose, and lignin are essential for the physical attributes of plant fibers, with cellulose serving as the primary organic element that imparts rigidity and strength to the fibers [58].

To improve the adherence of natural fibers to matrix materials, both physical and chemical treatments can alter the fiber surfaces by reducing hydrophilic hydrogen groups. Methods like surface fibrillation, cold plasma treatment, and electric discharge techniques adjust surface properties to promote stronger mechanical bonds with the matrix. Chemical treatment is another commonly utilized approach for modifying natural fiber surfaces, aiming to decrease their hydrophilic nature and increase compatibility with the matrix. Various chemical treatments, such as alkalization, acetalization, salinization, benzoylation, and peroxidation, have been applied to effectively modify fiber surfaces and enhance their adhesion and compatibility with matrix materials.

References

1. Peças P, Carvalho H, Salman H, Leite M (2018) Natural fibre composites and their applications: a review. J Compos Sci. https://doi.org/10.3390/jcs2040066
2. Sinha AK, Narang HK, Bhattacharya S (2017) Mechanical properties of natural fibre polymer composites. J Polym Eng 37:879–895. https://doi.org/10.1515/polyeng-2016-0362
3. Lu L, Fan S, Geng L et al (2021) Flow analysis of regenerated silk fibroin/cellulose nanofiber suspensions via a bioinspired microfluidic chip. Adv Mater Technol. https://doi.org/10.1002/admt.202100124
4. Ahmad F, Choi HS, Park MK (2015) A review: Natural fiber composites selection in view of mechanical, light weight, and economic properties. Macromol Mater Eng
5. Jaafar J, Siregar JP, Tezara C et al (2019) A review of important considerations in the compression molding process of short natural fiber composites. Int J Adv Manuf Technol 105:3437–3450. https://doi.org/10.1007/s00170-019-04466-8
6. Wang W, Guo X, Zhao D et al (2020) Water absorption and hygrothermal aging behavior of wood-polypropylene composites. Polymers (Basel) 12:782. https://doi.org/10.3390/polym12040782
7. Fan X, Keynton RS (2019) Fabrication and characterization of biopolymer fibers for 3D oriented microvascular structures. J Micromechanics Microengineering 29:083003. https://doi.org/10.1088/1361-6439/ab2349
8. Swolfs Y, Verpoest I, Gorbatikh L (2019) Recent advances in fibre-hybrid composites: materials selection, opportunities and applications. Int Mater Rev 64:181–215. https://doi.org/10.1080/09506608.2018.1467365
9. Babu* TN, Kuclourya T, Jain MK, Mangalaraja RV (2019) Mechanical behaviour of areca nut, sunn hemp natural fibers and E glass fibers reinforced with epoxy composites. Int J Eng Adv Technol 8:1092–1101. https://doi.org/10.35940/ijeat.F8322.088619
10. Chen W, Ji X, Huang Z (2021) Influence of fiber type on mechanical properties of lightweight cement-based composites. Sci Eng Compos Mater. https://doi.org/10.1515/secm-2021-0021
11. Pickering KL, Efendy MGA, Le TM (2016) A review of recent developments in natural fibre composites and their mechanical performance. Compos Part A Appl Sci Manuf 83:98–112. https://doi.org/10.1016/j.compositesa.2015.08.038
12. Bar M, Alagirusamy R, Das A (2018) Advances in natural fibre reinforced thermoplastic composite manufacturing: effect of interface and hybrid yarn structure on composite properties. In: Advances in natural fibre composites. Springer International Publishing, Cham, pp 99–117
13. Ali BAA, Salit MS, Zainudin ES, Othman M (2015) Integration of artificial neural network and expert system for material classification of natural fibre reinforced polymer composites. Am J Appl Sci. https://doi.org/10.3844/ajassp.2015.174.184
14. Nurazzi NM, Asyraf MRM, Khalina A et al (2021) A review on natural fiber reinforced polymer composite for bullet proof and ballistic applications. Polymers (Basel) 13:646. https://doi.org/10.3390/polym13040646
15. Kumar A, Biswal M, Mohanty S, Nayak SK (2021) Recent developments of lignocellulosic natural fiber reinforced hybrid thermosetting composites for high-end structural applications: a review. J Polym Res 28:459. https://doi.org/10.1007/s10965-021-02788-4

16. Soundhar A, Kandasamy J (2019) Mechanical, chemical and morphological analysis of crab shell/sisal natural fiber hybrid composites. J Nat Fibers. https://doi.org/10.1080/15440478.2019.1691127
17. Karthikeyan S, Rajini N, Jawaid M et al (2017) A review on tribological properties of natural fiber based sustainable hybrid composite. Proc Inst Mech Eng Part J J Eng Tribol 231:1616–1634. https://doi.org/10.1177/1350650117705261
18. Otto GP, Moisés MP, Carvalho G et al (2017) Mechanical properties of a polyurethane hybrid composite with natural lignocellulosic fibers. Compos Part B Eng. https://doi.org/10.1016/j.compositesb.2016.11.035
19. Gouda DAT, S P J, Dinesh DKR et al (2014) Characterization and investigation of mechanical properties of hybrid natural fiber polymer composite materials used as orthopaedic implants for femur bone prosthesis. IOSR J Mech Civ Eng 11:40–52.https://doi.org/10.9790/1684-11424052
20. Zeeshan M, Ali M, Anjum AS, Nawab Y (2021) Optimization of mechanical/thermal properties of glass/flax/waste cotton hybrid composite. J Ind Text. https://doi.org/10.1177/1528083719891420
21. Prabhu P, Karthikeyan B, Ravi Raja Malar Vannan R, Balaji A (2021) Dynamic mechanical analysis of Silk and Glass (S/G/S)/Pineapple and Glass (P/G/P)/Flax and Glass (F/G/F) reinforced Lannea coromandelica blender hybrid nano composites. J Mater Res Technolhttps://doi.org/10.1016/j.jmrt.2021.09.068
22. Chandgude S, Salunkhe S (2021) In state of art: Mechanical behavior of natural fiber-based hybrid polymeric composites for application of automobile components. Polym Compos 42:2678–2703. https://doi.org/10.1002/pc.26045
23. Safri SNA, Sultan MTH, Jawaid M, Jayakrishna K (2018) Impact behaviour of hybrid composites for structural applications: a review. Compos Part B Eng 133:112–121. https://doi.org/10.1016/j.compositesb.2017.09.008
24. Sathishkumar T, Naveen J, Satheeshkumar S (2014) Hybrid fiber reinforced polymer composites—a review. J Reinf Plast Compos 33:454–471. https://doi.org/10.1177/0731684413516393
25. Begum S, Fawzia S, Hashmi MSJ (2020) Polymer matrix composite with natural and synthetic fibres. Adv Mater Process Technol 6:547–564. https://doi.org/10.1080/2374068X.2020.1728645
26. Ravishankar B, Nayak SK, Kader MA (2019) Hybrid composites for automotive applications—a review. J Reinf Plast Compos. https://doi.org/10.1177/0731684419849708
27. Thakur VK, Thakur MK, Gupta RK (2014) Review: raw natural fiber-based polymer composites. Int J Polym Anal Charact 19:256–271. https://doi.org/10.1080/1023666X.2014.880016
28. Athijayamani A, Thiruchitrambalam M, Natarajan U, Pazhanivel B (2009) Effect of moisture absorption on the mechanical properties of randomly oriented natural fibers/polyester hybrid composite. Mater Sci Eng A 517:344–353. https://doi.org/10.1016/j.msea.2009.04.027
29. Panthapulakkal S, Raghunanan L, Sain M et al (2017) Natural fiber and hybrid fiber thermoplastic composites. In: Green composites. Elsevier, pp 39–72
30. Di BY, Joshi SC, Bhudolia SK, Gohel G (2020) Recent advances on the design automation for performance-optimized fiber reinforced polymer composite components. J Compos Sci 4:61. https://doi.org/10.3390/jcs4020061
31. Azman MA, Asyraf MRM, Khalina A et al (2021) Natural fiber reinforced composite material for product design: a short review. Polymers (Basel)
32. Lopes Lima PR, Roque AB, Ariani Fontes CM et al (2017) Potentialities of cement-based recycled materials reinforced with sisal fibers as a filler component of precast concrete slabs. In: Sustainable and nonconventional construction materials using inorganic bonded fiber composites. Elsevier, pp 399–428
33. Pandey P, Bajwa D, Ulven C, Bajwa S (2016) Influence of hybridizing flax and hemp-agave fibers with glass fiber as reinforcement in a polyurethane composite. Materials (Basel) 9:390. https://doi.org/10.3390/ma9050390

34. Maslinda AB, Abdul Majid MS, Ridzuan MJM et al (2017) Effect of water absorption on the mechanical properties of hybrid interwoven cellulosic-cellulosic fibre reinforced epoxy composites. Compos Struct. https://doi.org/10.1016/j.compstruct.2017.02.023
35. Shahria S (2019) Fabrication and property evaluation of hemp–flax fiber reinforced hybrid composite. Chem Mater Eng 7:17–23. https://doi.org/10.13189/cme.2019.070202
36. Pan Y, Zhong Z (2015) The effect of hybridization on moisture absorption and mechanical degradation of natural fiber composites: an analytical approach. Compos Sci Technol 110:132–137. https://doi.org/10.1016/j.compscitech.2015.02.005
37. Furtado SCR, Araújo AL, Silva A et al (2014) Natural fibre-reinforced composite parts for automotive applications. Int J Automot Compos 1:18. https://doi.org/10.1504/IJAUTOC.2014.064112
38. Nurazzi NM, Asyraf MRM, Fatimah Athiyah S et al (2021) A review on mechanical performance of hybrid natural fiber polymer composites for structural applications. Polymers (Basel) 13:2170. https://doi.org/10.3390/polym13132170
39. Saravana Bavan D, Mohan Kumar GC (2010) Potential use of natural fiber composite materials in India. J Reinf Plast Compos. https://doi.org/10.1177/0731684410381151
40. Lee CH, Padzil FNBM, Lee SH et al (2021) Potential for natural fiber reinforcement in pla polymer filaments for fused deposition modeling (Fdm) additive manufacturing: a review. Polymers (Basel)
41. Shahinur S, Alamgir Sayeed MM, Hasan M et al (2022) Current development and future perspective on natural jute fibers and their biocomposites. Polymers (Basel)
42. Holbery J, Houston D (2006) Natural-fiber-reinforced polymer composites in automotive applications. JOM
43. Arun Prakash VR, Viswanthan R (2019) Fabrication and characterization of echinoidea spike particles and kenaf natural fibre-reinforced Azadirachta-Indica blended epoxy multi-hybrid bio composite. Compos Part A Appl Sci Manuf 118:317–326. https://doi.org/10.1016/j.compositesa.2019.01.008
44. Aravindh M, Sathish S, Prabhu L et al (2022) Effect of various factors on plant fibre-reinforced composites with nanofillers and its industrial applications: a critical review. J Nanomater. https://doi.org/10.1155/2022/4455106
45. Kumar N, Mehta V, Kumar S et al (2022) Bamboo natural fiber and PAN fiber used as a reinforced brake friction material: developed asbestos-free brake pads. Polym Compos. https://doi.org/10.1002/pc.26584
46. Koronis G, Silva A, Fontul M (2013) Erratum to "“Corrigendum to ‘Green composites: a review of adequate materials for automotive applications’ [Composites Part B: Engineering 44 (2013) 120–127]”" [Composites Part B: Engineering 47 (2013) 391]. Compos Part B Eng 49:93. https://doi.org/10.1016/j.compositesb.2013.02.011
47. Mohd Bakhori SN, Hassan MZ, Mohd Bakhori N et al (2022) Physical, mechanical and perforation resistance of natural-synthetic fiber interply laminate hybrid composites. Polymers (Basel)
48. Paladugu SRM, Divya Aparna N, Rama Sreekanth PS (2022) Experimental investigations and mechanical analysis of hybrid natural fibre reinforced composites. Mater Today Proc 56:1115–1119. https://doi.org/10.1016/j.matpr.2021.10.272
49. Akil HM, Santulli C, Sarasini F et al (2014) Environmental effects on the mechanical behaviour of pultruded jute/glass fibre-reinforced polyester hybrid composites. Compos Sci Technol 94:62–70. https://doi.org/10.1016/j.compscitech.2014.01.017
50. Zamri MH, Akil HM, Bakar AA et al (2012) Effect of water absorption on pultruded jute/glass fiber-reinforced unsaturated polyester hybrid composites. J Compos Mater. https://doi.org/10.1177/0021998311410488
51. Sergi C, Tirillò J, Seghini MC et al (2019) Durability of basalt/hemp hybrid thermoplastic composites. Polymers (Basel). https://doi.org/10.3390/polym11040603
52. Todkar SS, Patil SA (2019) Review on mechanical properties evaluation of pineapple leaf fibre (PALF) reinforced polymer composites. Compos Part B Eng

53. Ferrante L, Tirillò J, Sarasini F et al (2015) Behaviour of woven hybrid basalt-carbon/epoxy composites subjected to laser shock wave testing: Preliminary results. Compos Part B Eng. https://doi.org/10.1016/j.compositesb.2015.03.084
54. Fiore V, Scalici T, Calabrese L et al (2016) Effect of external basalt layers on durability behaviour of flax reinforced composites. Compos Part B Eng. https://doi.org/10.1016/j.compositesb.2015.08.087
55. Amiandamhen SO, Meincken M, Tyhoda L (2018) The effect of chemical treatments of natural fibres on the properties of phosphate-bonded composite products. Wood Sci Technol. https://doi.org/10.1007/s00226-018-0999-9
56. Hossain SI, Hasan M, Hasan MN, Hassan A (2013) Effect of chemical treatment on physical, mechanical and thermal properties of ladies finger natural fiber. Adv Mater Sci Eng 2013:1–6. https://doi.org/10.1155/2013/824274
57. Aravindh M, Sathish S, Ranga Raj R et al (2022) A review on the effect of various chemical treatments on the mechanical properties of renewable fiber-reinforced composites. Adv Mater Sci Eng
58. Fiore V, Scalici T, Nicoletti F et al (2016) A new eco-friendly chemical treatment of natural fibres: effect of sodium bicarbonate on properties of sisal fibre and its epoxy composites. Compos Part B Eng. https://doi.org/10.1016/j.compositesb.2015.09.028

Chapter 2
Review of Hybrid Fiber Based Composites

Abstract This chapter provides an exhaustive examination of the characteristics of various natural and synthetic fibers utilized in the formation of pure natural composites as well as hybrid composites predicated on the amalgamation of both natural and synthetic fibers. It further addresses bio-based resins, a variety of fabrication techniques, the chemical and mechanical properties of fibers, and the impacts of chemical treatment alongside the influence of nanoparticles on composite materials (Atmakuri et al. in Polym 12(9):1–30, 2020 [1]). Natural fibers have garnered increasing interest and esteem among researchers, delivering satisfactory results due to their abundant availability, simplicity of fabrication, cost-effectiveness, biodegradability, and environmental sustainability. Hybrid composites, composed of two distinct natural fibers within the same matrix material, are more prevalent than combinations of natural and synthetic fibers. Contemporary studies concerning natural fiber hybrid composites have demonstrated that, in light of their biodegradability and the influence of the individual fibers' strength on mechanical properties, flame retardancy, and moisture absorption, natural fibers necessitate additional treatments, such as chemical treatment, to overcome these limitations and enhance their properties. The effect of chemical treatment on the thermal, mechanical, and moisture characteristics of composite materials has been investigated. Researchers have identified a favorable impact on overall strength resulting from the incorporation of filler materials (nanoparticles) in composite materials. Hybrid composites represent a central topic in polymer science, attracting significant attention for a range of lightweight applications across a wide array of industries, including the automotive, construction, shipping, aviation, sports equipment, electronics, hardware, and biomedical sectors.

2.1 Introduction

In recent years, there has been a growing interest among researchers in the application of natural fibers for the development of hybrid composites, attributed to their abundance, low density, recyclability, and significant strength [2]. Natural fibers have garnered considerable attention as a viable alternative to synthetic fibers [3],

A. Atmakuri et al., *Fiber-Based Hybrid Composites for Engineering*, Springer Series on Polymer and Composite Materials, https://doi.org/10.1007/978-3-031-93144-4_2

particularly as a reinforcement medium for various innovative applications. This interest stems from their inherent properties, including natural availability, cost-effectiveness, and environmental friendliness. These fibers do not emit harmful gases and actively absorb carbon dioxide, thereby contributing to the reduction of environmental pollution [4–6].

One of the primary limitations of employing natural fibers as reinforcement in composites is their hydrophilic nature, which renders them incompatible with hydrophobic matrices. To ameliorate this inconsistency, surface treatment of natural fibers is necessary to enhance the overall properties of the composite [7, 8]. Furthermore, natural fibers are highly flammable; thus, to mitigate this drawback, various fire retardants have been incorporated to enhance their fire resistance. Surface modifications involving various chemical compounds have been shown to improve their mechanical and physical properties. Additionally, structural modification of composite materials leads to increased stability [9, 10]. Over recent years, there has been a growing interest in integrating at least two fillers into a common matrix. The past decade has witnessed a steady increase in the number of publications pertaining to hybrid composites. This trend is reflected in the analysis of hybrid composites (from 2010 to 2020) in Web of Science and Scopus, where the term 'hybrid composites' yielded 26,450 research outputs. This development stems from a singular objective: to enhance the properties of the resulting composite material to attain the desired characteristics and improved performance of such composites.

Figure 2.1 illustrates the aggregate research findings on hybrid composites over recent years, with data sourced from the Web of Science. A majority of the research publications originate from China, followed by India. The hybridization process is extensively represented in research articles across various fields, including multidisciplinary materials, chemical industry, polymer science, nanoscience, nanotechnology, materials science and engineering, electrochemistry, mechanics, and mechanical engineering, among others.

Figure 2.2 illustrates the aggregate research publications on natural fiber hybrid composites. The data was obtained from the Web of Science through a search for "natural fibers and hybrid composites." A total of 1460 research outputs were identified, with a majority of contributors hailing from the fields of material science and polymeric science engineering. The commencement of the hybridization processes of natural fiber composites is traced back to early 2008, with a subsequent gradual increase in scholarly interest. The primary objective is to enhance composite properties through the incorporation of two or three different filler materials within the same matrix material. Numerous researchers have demonstrated that the orientation of filler materials significantly influences the composite [11–13].

Hybrid Composites consist of two or more constituent materials, specifically matrix and reinforcement [14, 15]. The matrix serves as a coupling material that binds the reinforcement. Reinforcements are materials embedded within the matrix, providing additional strength to the composite. The importance of composites lies in their ability to offer enhanced physical, chemical, and mechanical properties, distinct from their individual constituents [16]. Composites are categorized based on the matrix type into polymer, ceramic, and polymer. Variations based on reinforcement

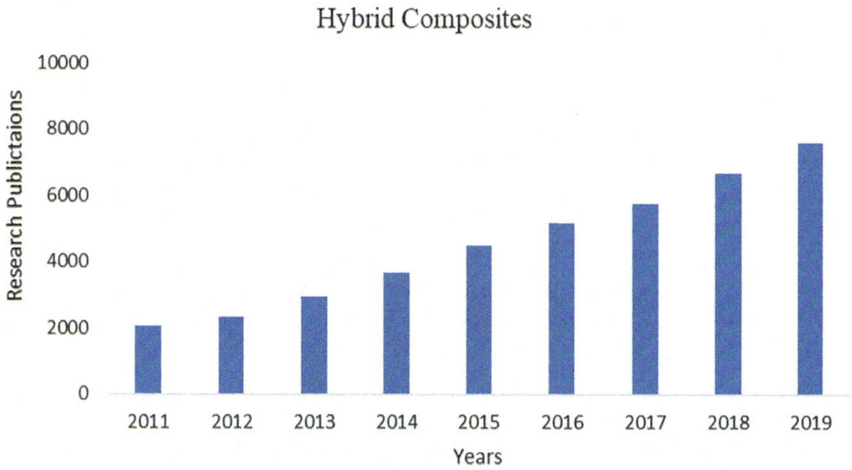

Fig. 2.1 Recent publications on hybrid composites

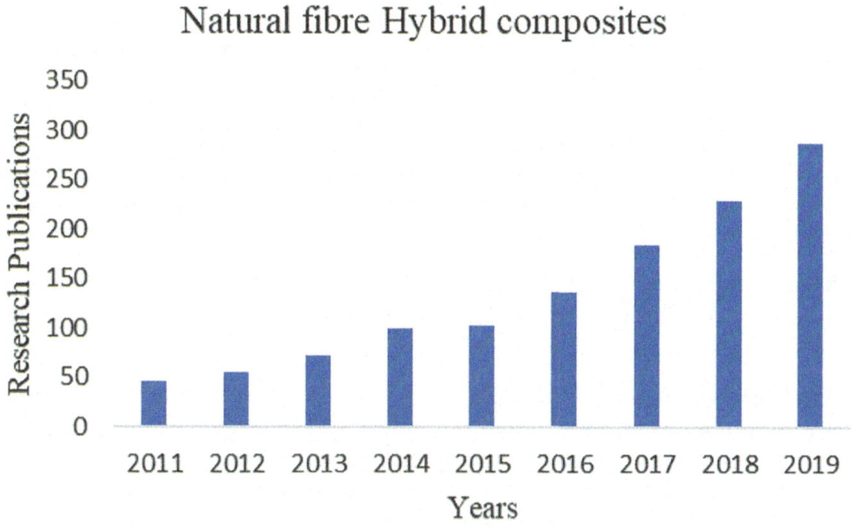

Fig. 2.2 Recent publications on natural fiber hybrid composites

include laminate, fibrous, and particulate composites. Reinforcements are classified, in terms of availability, as either natural or synthetic [15, 17]. When the compound material comprises metal fiber, it is termed a metal matrix composite; if it involves a polymer matrix, it is referred to as a polymer matrix material [17, 18]. Fiber-reinforced polymers (FRP) consist of a polymer matrix, coupled with high-quality fibers clear intersections between them, thus maintaining their chemical and physical properties.

2.2 Classification of Hybrid Fiber Composites

Natural hybrid composite materials consist of high-quality fibers, integrated within a matrix material to enhance modulus. Consequently, each individual component (fiber and matrix) retains its inherent chemical and physical characteristics, resulting in a synergy of properties unattainable if utilized independently. In this context, the fibers serve as the primary load-bearing elements, while the matrix material maintains the integrity of their position, facilitating stress distribution among them and safeguarding the fibers against environmental damages such as elevated temperature and humidity. Accordingly, it is proper to assert that fibers provide reinforcement to the matrix; hence the term Fiber Reinforced Composite Materials (FRCM). Reinforced hybrid materials are constructed by amalgamating at least two distinct types of fibers within a unique matrix material. The hybridization of two different types of fibers, characterized by specific lengths and widths, offers several advantages over the utilization of either fiber alone within a single polymer matrix. Predominately, research focuses on the hybridization of natural fibers with glass fibers to enhance their properties. These composite materials present average calorific values and impose minimal concerns regarding safety, well-being, and protection. Additionally, they demonstrate exceptional mechanical properties, possess low density, and are economically viable [7, 14, 19–21]. The superiority of composite materials over traditional materials, such as enhanced rigidity and durability, renders them more versatile in application. The accessibility and cost-effectiveness of these natural fibers intrigue researchers, yielding promising outcomes in the fabrication of composites for practical applications. Composite materials are composed of at least two materials with physically separable phases.

Characterisation of Composites. Composite materials classified on the basis of a matrix serve as a binder and a reinforcement, which acts as a filler material. Figure 2.3 shows the classification of composites.

Metal Matrix Composites (MMCs). Metal matrix composites offer several advantages over traditional solid metals, including superior strength, modulus, and enhanced properties at elevated temperatures. Owing to these attributes, metal matrix composites are being considered for a broad range of applications. However, due to their high density, they do not experience significant demand. The most commonly utilized metals in these composites are steel, tungsten, boric, and molybdenum [22, 23].

Ceramic Matrix Composites (CMCs). Ceramic matrix composites and carbon-based composites exemplify inorganic (non-metallic) composite materials. The primary objective of CMCs is to enhance the toughness of composite materials [24] and they are designed to address the limitations associated with monolithic ceramics. The challenges posed by matrix materials are less significant compared to the failure of fiber materials, thereby exerting a considerable influence on composite fabrication. To circumvent typical processing drawbacks (premature failure), several

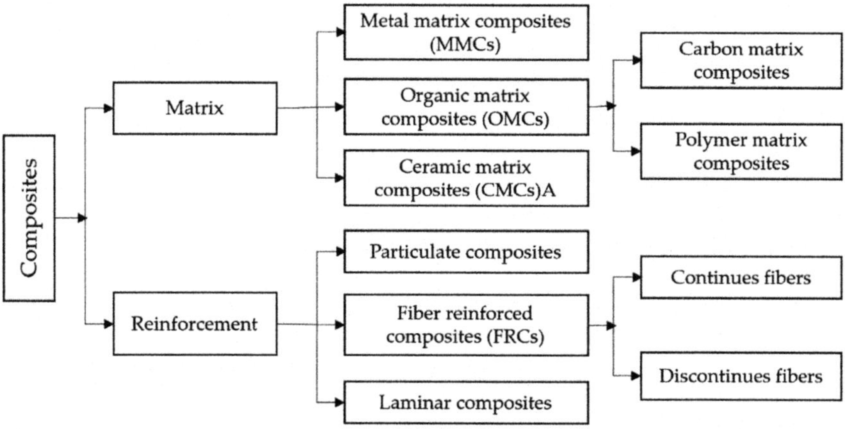

Fig. 2.3 Classification of composites [22]

methodologies are employed, including gas-phase and liquid-phase fabrication of CMCs.

Organic Matrix Composites (OMCs). Organic matrix composites can be categorized into polymer matrix composites and carbon–carbon matrix composites. Among these, polymer composites are the most widely utilized matrix materials. The mechanical properties of polymeric composites lend themselves effectively to diverse structural applications due to their advantageous attributes, including low mass, high stiffness, and resistance to elevated temperatures. In contrast to metal and ceramic composites, polymer composites exhibit lower strength and stiffness; this limitation is addressed by reinforcing the composites with alternative materials. The process of handling polymer composites does not necessitate high temperature and pressure. Similarly, the equipment necessary for the fabrication of composites is less complex. Consequently, polymer matrix composites gained rapid recognition and became a focal point of scholarly interest due to their fundamental applications. Composites are employed because their overall properties surpass those of individual components. Polymer composites are further subdivided into thermoset and thermoplastic composites. Thermoplastic materials exist as solids at ambient temperature; they require heating to soften or melt for shaping into molds, followed by curing to achieve the desired form. These materials are recyclable. In contrast, thermosetting polymers are available in both liquid and solid states, yet are non-recyclable. In thermoplastics, atoms are connected via covalent bonds, while thermosetting polymers feature cross-linked bonds, rendering them stronger and more rigid than thermoplastics. Thermoplastic materials encompass polyamides, polyoxymethylene (POM), polycarbonates (PC), and acrylics. Thermosetting materials include epoxies (EP), thermoset polyamides, amino plastics, and polyester thermosets (TS) [24].

Fiber Reinforced Polymer Composites (FRCs). Fiber Reinforced Polymer Composites represent the most extensively utilized composites, consisting of fiber and matrix elements. The fiber functions as a reinforcement and contributes to the stiffness of the composite material. The matrix component binds all fibers cohesively and facilitates the distribution of stress among the fibers. Occasionally, filler materials are incorporated to enhance the fabrication process. Based on their availability, reinforcements are categorized into natural and synthetic fibers. Despite the numerous advantages associated with synthetic composites, certain drawbacks exist. Synthetic fibers exhibit a higher propensity to combust in comparison to natural fibers. Additionally, rubbing synthetic fibers with natural fibers generates additional electrostatic charge [23, 25].

2.3 Classification of Fibers

Natural Fibers. Natural fibers are derived from animal, plant, and mineral-based sources, where the primary components are protein, cellulose, and minerals, respectively. These fibers exhibit superior physical and mechanical properties in comparison to alternative fibers. They are abundantly available in nature and possess biodegradability [26, 27]. Due to their lightweight, high strength, cost-effectiveness, and resistance to corrosion, natural fibers are increasingly being utilized in industrial applications [28]. However, they present certain drawbacks, such as high water absorption, significant anisotropy, reduced homogeneity relative to synthetic fibers, and stress development within composite materials [28]. Figure 2.4 illustrates the classification of natural fibers.

The chemical composition is a crucial determinant in fiber materials and is contingent upon the fiber source. It exhibits variability across different fibers and within their respective families, as illustrated in Table 2.1. The performance of these fibers is predominantly reliant on the aspect ratio and the crystallinity of cellulose within the fibers. Previous research has indicated that these fibers possess biodegradability due to their chemical constituents. For instance, UV degradation attributable to lignin and

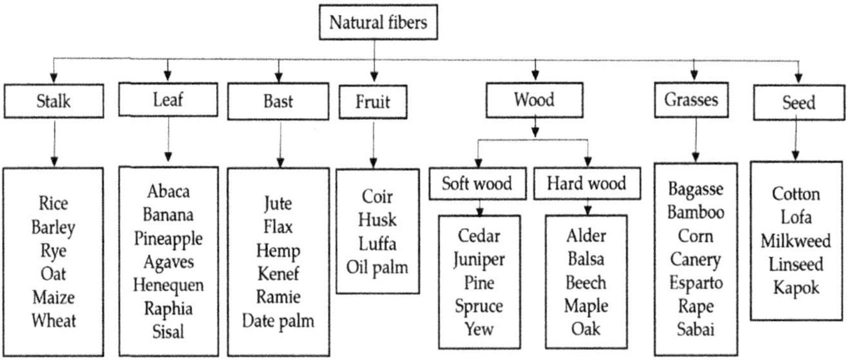

Fig. 2.4 Classification of natural fibers [22–24]

hemicellulose is responsible for both thermal and biodegradation, as well as water absorption. The physical and mechanical properties of natural fibers are presented in Table 2.2. The primary components of natural fibers are cellulose, hemicellulose, and lignin. Due to the developmental conditions, the proportions of these components vary accordingly. To examine the behavior of these fibers, it is imperative to consider the structural composition of these constituents. Cellulose serves as a principal building block of natural fibers, constituting approximately 50% of plant materials. The presence of cellulose within plant fibers significantly influences moisture absorption. Hemicellulose, the second major component following cellulose, is less significant as a filler material. Lignin is comprised of phenylpropane derivatives and is an amorphous polymeric material that regulates fluid transport and cohesively binds the rigid cells within the plant. Lignin functions as a cementing material, thereby affecting the structure, properties, and morphology [29]. Some of the advantages and disadvantages of natural fibers are elaborated in Table 2.3.

Synthetic Fibers. Numerous artificially produced or synthetic fibers have recently become the focus of research networks due to their exceptional properties. Table 2.5 presents the mechanical properties of synthetic fibers. Among all synthetic fibers, glass and carbon fibers are predominantly employed as filler materials in hybrid composites. Production of these fibers involves the concepts of polymerization,

Table 2.1 Natural fibers chemical composition [22, 24–26]

Fiber type	Cellulose (%)	Hemi-Cellulose (%)	Lignin (%)	Pectin (%)	Wax (%)	Ash (%)	Moisture (%)
Abaca	56–63	20–25	7–9	–	3	–	–
Bamboo	26–43	30	1–31	–	10	–	9.16
Banana	63–83	–	5	–	11	–	10.71
Coir	36–43	41–45	0.15–0.25	3–4	–	–	11.36
Cotton	83–91	3	–	0.6	8–9	–	7.85–8.5
Flax	64–72	64–72	2–2.2	1.8–2.3	–	–	8–12
Hemp	70–74	0.9	3.7–5.7	0.8	1.2–6.2	0.8	6.2–12
Jute	61–72	18–22	12–13	0.2	0.5	0.5–2	12.5–13.7
Kenaf	45–57	8–13	22	0.6	0.8	2–5	6.2–12
Nettle	86	5.4	4.0	0.6	3.1	–	–
Rachis	43	–	26	–	–	–	–
Ramie	69–91	5–15	0.4–0.7	1.9	–	–	–
Rice husk	38–45	–	–	–	–	20	–
Sisal	78	10	8	–	2	1	11
Hardwood	38–49	17–23	23–30	–	–	–	–
Softwood	40–45	19–22	26–34	–	–	–	–

Table 2.2 Natural fibers physical and mechanical properties [22, 26–28]

Fiber type	Density (g/cm^3)	Tensile Strength (MPa)	Young's Modulus (GPa)	Elongation (%)	Specific strength (S/ρ)	Specific Modulus (E/ρ)
Abaca	1.5	430–813	31.1–33.6	2.9–10	267	8
Bagasse	0.89	350	22	5.5	–	–
Bamboo	0.6–1.25	290	11–17	–	454	32.6
Banana	1.35	529–914	27–32	2.6–5.9	444	13.2
Coir	1.15–1.25	131–220	4–6	15–40	146	3.3–5
Cotton	1.51	400	12	3–10	179–373	3.44–7.9
Flax	1.4	800–1500	60–80	1.2–1.6	535–1000	18.4–53
Hemp	1.48	550–900	70	1.6–4.0	372–608	47.3
Jute	1.31–1.48	393–800	13–26.5	1.16–1.8	269–548	6.85–20.6
Kenaf	1.4	284–930	21–60	1.6	641	36.55
Nettle	1.51	650	38	1.7	–	87–115
Pineapple	1.44	413–1627	60–82	14.5	–	–
Ramie	1.5	500	44	2	147–675	29.3–85
Sisal	1.3–1.4	390–450	12–41	2.3–2.5	366–441	6.5–15.2
Softwood	1500	1000	40	–	667	26.67
Hardwood	1200	950	37.9	–	–	–

Table 2.3 Merits and demerits of natural fibers

Merits	Demerits
Lightweight	Flammable
Recyclable	Dimensional instability
Improved specific mechanical properties	High moisture absorption
Eco-friendly	Anisotropic behavior
Do not generate any harmful gasses during processing	Limited processing temperature (~ 200 to 230 °C)
Good thermal properties	Sensitive to UV
Good acoustic properties	Fugal attack and microbial
Low cost, availability	Low strength than synthetic fibers

which entail the linking of monomers to create an extended chain. Table 2.4 elucidates the advantages and disadvantages of synthetic fibers. Figure 2.5 illustrates the classification of synthetic fibers.

Table 2.4 Advantages and disadvantages of synthetic fibers

Advantages	Disadvantages
Long lasting	Flammable
Readily pick-up to various dyes	Prone to heat damage
Stretchable	Melt easily
Waterproofing	Not eco-friendly
Non biodegradability	Cause for microplastic pollution
Moisture resistance	Not suitable for hot washing
Strain and wear resistance	Poor insulation capacity

Table 2.5 Mechanical properties of synthetic fibers [26–28]

Type of fiber	Density (g/cm^3)	Tensile strength (MPa)	Young's modulus (GPa)	Elongation (%)
Aramid	1.44	3000	124	2.5
Carbon	1.4	400	230–240	1.4–1.8
E-Glass	2.55	2000–3500	63–67	2.5
S-Glass	2.5	4570	70–73	1.8–3.2
Silicon carbide	3.16	–	360–440	–

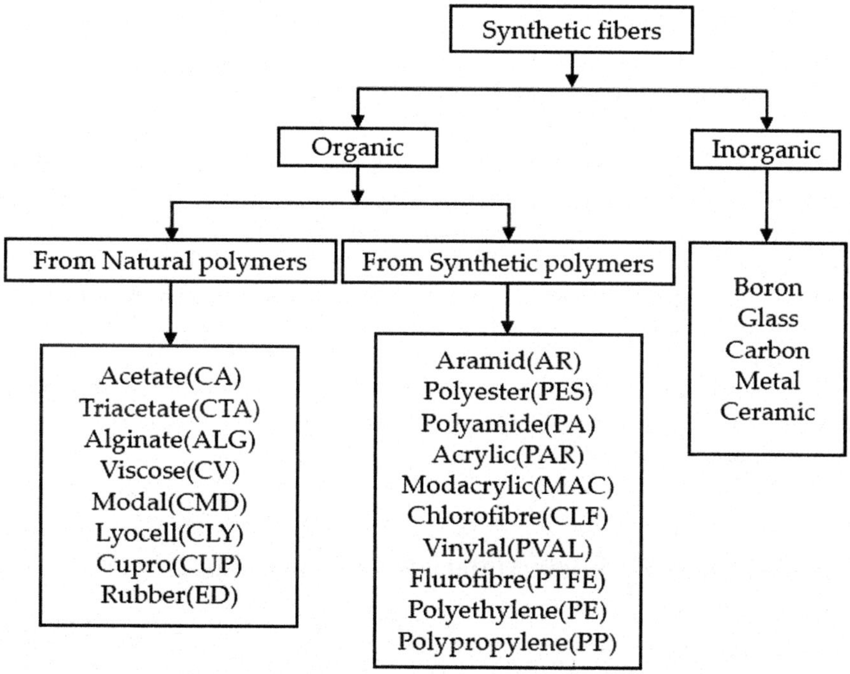

Fig. 2.5 Classification of synthetic fibers [15]

Table 2.6 Chemical treatment processes and its importance on hybrid fiber composites

Chemical Treatment	Fibers	Advantages	References
Alkaline treatment	Flax Jute Sisal	Improves the rough surface Better mechanical interlocking, and cellulose content	[32–34]
Silane treatment	Glass	Prevent swelling of composite Improve thermal stability	[35, 36]
Acetylation treatment	Sisal	Improve the bio-resistance of composite and fiber-matrix interactions	[37, 38]
Benzoylation treatment	Flax Sisal	Reduces the moisture absorption, improve its thermal stability and strength	[39–41]
Acrylonitrile grafting and Acrylation	Oil palm Glass	It improves the tensile strength and young's modulus	[42]
Malleated coupling gents	Cellulose fibers Banana Hemp	Growth in flexural strength, hardness, impact strength and flexural modulus	[43, 44]
Permanganate treatment	Cellulose fibers	It improves the hydrophilic tendency of the composite	[45]
Peroxide treatment	Oil palm	Improves tensile properties	[45]
Isocyanate treatment	Leaf fibers	Enhance better mechanical properties	[46, 47]

2.4 Chemical Treatment of Natural Fibers

Natural fibers, despite possessing exceptional qualities in comparison to other advanced fibers, are not exempt from defects. The presence of strong polar bonds within these fibers can potentially hinder their bonding efficacy with the polymer matrix [30]. To mitigate this issue, the implementation of chemical treatments is necessary to enhance the bonding capability. Such treatments contribute additional strength to the composite material, facilitating a robust interaction between fiber and matrix material during the curing process. Furthermore, chemical treatments enhance the surface quality and reduce the moisture content, i.e., the amount of water absorbed by the composite material [31]. A selection of these treatment techniques is enumerated in Table 2.6.

2.5 Selection of Matrix Materials for Fabrication

A matrix material is integral to composite materials as it significantly influences their characteristics. The morphological traits, durability, and surface properties are contingent upon the matrix (resin) material. According to processing techniques, resin materials are categorized into two classifications: thermosets and thermoplastics [48]. The matrix serves as a load distributor across the reinforcement material when subjected to external pressure. Thermoplastics are more prevalently utilized than thermosets due to their moldability, as they readily assume the desired shape within molds. Conversely, thermosets are not recyclable because they do not revert to their original state upon transitioning from liquid to solid during the curing process. Thermoplastics, however, can be reshaped through the application of heat.

Thermoplastics. Researchers have demonstrated increased interest in employing thermoplastics as matrix materials due to their essential properties, including low cost, moldability, reduced weight, and recyclability. The principal advantages and disadvantages of these thermoplastic polymer materials are elucidated in Table 2.7. The viscosity range exceeds that of uncured thermosets by more than 500 times during the melting process. Recent findings have established that the application of chemical treatments to natural fibers may enhance the bonding properties between matrix and reinforcement [49], with treatments including alkaline, silane, acetylation, and benzoylation, among others. The incorporation of chemical and biological treatments consequently enhances the overall properties of composites [50]. Among synthetic polymers, thermoplastics are the most commercially utilized materials. The most extensively used thermoplastics include polyvinyl chloride (PVC), polystyrene (PS), polypropylene (PP), polyethylene, and natural rubber. The mechanical properties of thermoplastic resins are detailed in Table 2.8. Kasama et al. examined the mechanical and thermal properties of sisal and glass fibers as reinforcements, with polypropylene serving as the matrix material [51]. The authors noted that integrating polypropylene into hybrid composites results in enhanced thermal stability, moisture resistance, and improved interfacial adhesion between reinforcement and matrix. Schmidt et al. detailed the mechanical properties of hybrid composites applying the RTM (resin transfer molding) technique [52]. They evaluated both woven and non-woven composites by assessing the in-plane permeability of the hybrid composites, contrasting these for comparison purposes, and reported enhanced mechanical properties and flow media of sisal fibers and glass mats with the inclusion of PP in the hybrid composite. Nayak et al. summarized the dynamic mechanical and rheological properties of bamboo glass fiber hybrid composite PP as a resin material using an injection molding technique [53]. Notably, the researcher observed that substituting Maleic anhydride grafted polypropylene (MAPP) for standard PP yielded superior results. Vega-Hernandez et al. developed composites based on blue agave bagasse utilizing polystyrene as a resin matrix [54]. They investigated the mechanical performance of the hybrid composites through in situ reversible addition-fragmentation chain transfer (RAFT).

Haneefa et al. conducted research on the tensile and flexural properties of banana and glass fiber reinforced polystyrene composites [55]. It was found that the rigidity and modulus of the composite increase with increments in the volume fraction of glass fiber. This observation is attributed to the superior compatibility of glass fiber with polystyrene compared to banana fiber. In contrast, the elongation at break

Table 2.7 Thermoplastic resins advantages and disadvantages

Type of resin	Advantages	Disadvantages
Polystyrene (PS)	Moisture resistance Weather resistance Good chemical resistance	Flammable Brittle Low impact resistance
Polypropylene (PP)	Good fatigue resistance High thermal resistance High dielectric resistance Excellent chemical resistance	Hard to process Comparatively high cost Limited availability
Polyethylene (PE)	Low cost Lightweight Low moisture absorption Good fatigue resistance Good impact strength High ductility	Flammable High thermal expansion Poor weather resistance
Polyvinyl chloride (PVC)	Versatility High impact strength Fire retardant Advance in Chemical Resistance Less cost Excellent dimensional stability	Poor resistance to UV and temperature
Natural rubber	Good tensile strength High resilience Tear resistance Low cost Water resistance	Poor resistance to hydrocarbons Chances for swelling Complete dissolution

decreases with increased volume fractions of glass fibers due to the lower elongation at break characteristic of glass fiber. Experimental findings demonstrated that chemically modified composites exhibit mechanical properties that are 30% greater than those of conventional composites. Hao X et al. investigated the effects of fiber geometry and alignment on the anisotropy of the flexural and thermal expansion properties of natural fiber polyethylene hybrid composites [56]. Samples were manufactured through the extrusion process, and to assess fiber orientation effectiveness, composites were cut at various offset angles for comparison. The findings revealed that composites oriented at zero degrees exhibited the optimal mechanical properties, while some deviation was observed at a 90-degree orientation. Robledo-Ortiz et al. developed green polyethylene natural fiber biocomposites utilizing a rotational molding technique [57]. In this study, biocomposites were produced using the rotational molding (rotomolding) technique. Natural fibers were integrated into a low-density green polyethylene matrix, and post-fabrication, samples underwent testing for mechanical properties. Maleated polyethylene (MAPE) was employed as a surface treatment to enhance adhesion properties. The results suggested that the less elastic nature was due to inadequate bonding capacity between the reinforcement and the matrix. Paran et al. explored the dynamic properties of green thermoplastic elastomer vulcanizate (GTEV) nanocomposites based on polyvinyl chloride

Table 2.8 Mechanical properties of thermoplastic resins

Polymer	Density (g/cm^3)	Tensile strength (MPa)	Young's modulus (GPa)	Elongation (%)	Glass Temp. (°C)
Polyethylene (PE)	0.9	18.0	0.5	350.0	− 78.0
High-density polyethylene (HDPE)	0.9–1.0	32.0–38.2	1.3	150.0	− 110.0
Low-density polyethylene (LDPE)	0.9	10.0–11.6	0.2–0.3	400.0	− 110.0
Polyethylene terephthalate (PET)	1.5–1.6	55.0–159.0	2.3–9.0	300.0	67.0
Polystyrene (PS)	1.04	34.0	3.0	1.6	100.0
High-impact polystyrene (HIPS)	1.0	42.0	2.1	2.5	88.0–92.0
Polycarbonate (PC)	1.2	69.0	2.3	200.0	147.0
Polypropylene (PP)	0.9–1.3	35.8	1.6	80.0	− 20.0
Poly methyl methacrylate (PMMA)	1.1–1.2	72.4	3.0	2.5	125.0
Polyvinyl chloride (PVC)	1.3–1.5	52.0–90.0	3.0–4.0	50.0–80.0	82.0
Nylon 6 (PA 6)	1.1	81.4	2.8	60.0	47.0
Nylon 6,6 (PA 6,6)	1.1	82.7	2.8	60.0	70.0

(PVC) and rice straw as reinforcements [58]. Surface morphological studies revealed that the GTEV nanocomposite exhibited significant surface roughness and fracture, indicating certain interactions between the matrix material and natural fibers [59]. The effect of fiber loading was predicted through the application of parallel and series equations [60]. Boussehel et al. examined the impact of chemical treatment on polyvinyl chloride natural fiber hybrid composites [61]. In this investigation, the authors compared PVC composites with and without chemical treatment, revealing a slight variation in mechanical properties [62]. However, chemical treatment induced structural changes in natural fibers due to the formation of hydroxyl bonds [63]. Haseena et al. conducted a study on the interfacial adhesion of natural hybrid fiber natural rubber composites using the equilibrium technique [64]. The swelling coefficient was observed to decrease with an increase in fiber content and penetrant size.

The addition of bonding agents resulted in greater resistance to expansion due to stronger interfacial adhesion and consequently higher cross-linked density [65]. The recent research on various fabrication techniques of thermoplastic resins is discussed in Table 2.9.

Thermosets Thermosets are renowned for their distinctive capability to form three-dimensional bonds post-curing. Thermoset resins differ from thermoplastics, possessing enhanced mechanical properties, thermal stability, resistance to creep, and biodegradability [83]. The advantages and disadvantages of various thermoset resins are detailed in Table 2.10. Unlike thermoplastics, thermosets do not allow for reuse [84]. Examples of thermoset polymers include epoxy resin, vinyl ester, polyester, phenolic, and urea–formaldehyde, among others. Among these, epoxy and vinyl ester resins are predominantly utilized due to their accessibility, flexibility, and cost-effectiveness [85]. A primary challenge associated with thermosets is the extended time required for handling. Furthermore, given the low recyclability of these composites, there is a need for innovative methods to develop sustainable recycling components. For instance, they can be crushed and reused as filler materials. The main objective has been to enhance the durability of thermosets and optimize handling temperatures. Owing to their superior mechanical properties, as demonstrated in Table 2.11, thermoset-based composites are in significant demand for commercial applications such as the aerospace industry, sports equipment, lightweight objectives, construction, automotive, biomedical, and numerous other practical uses. Among these, epoxy resin, in conjunction with a hardener, is predominantly employed in the fabrication of hybrid composites. Various processing techniques (Table 2.12) are currently available, including the hand lay-up method, the combination of hand lay-up with compression molding, pultrusion, hot press molding, twin-screw extrusion, and the vacuum-assisted resin transfer molding technique (VARTM). The hand lay-up process is predominantly employed due to its cost-effectiveness, simplicity, and improved outcomes. Although there are certain drawbacks, such as the curing process, non-uniformity of resin material, and vacuum voids within the composite, these are significantly reduced compared to other techniques.

Yorseng et al. developed bio-based natural fiber hybrid bio-epoxy composites for applications in semi-structured components, observing their mechanical behavior under various conditions [86]. Surface morphology was investigated using contact angle measurement properties to determine the wettability of the composites. The results indicated a reduction in mechanical properties due to bio-epoxy degradation under accelerated weathering conditions, and an increase in the ductile nature of the composite leading to breakdown [87–89]. Thermal studies demonstrated that bio-epoxy based natural fiber composites can withstand temperatures up to 295 °C, rendering them suitable for high-temperature applications. Alves et al. examined the dynamic mechanical properties of hybridized jute and biaxial glass fiber reinforcement with epoxy as the matrix material in composites [90]. The vacuum infusion technique was employed to achieve optimal impregnation levels for both fibers during the fabrication process. This paper presents woven and non-woven fabric reinforced epoxy resin composites and evaluated the impact of fiber arrangement on

Table 2.9 Recent research on thermoplastic hybrid composites and fabrication techniques

Matrix	Reinforcement	Fabrication process	Properties observed	References
Polyethylene (PE)	Coir fibers	Rotational molding	Water absorption SEM Mechanical measurements	[66]
Polyethylene (PE)	Cupula fibers	Screw extruder	Tensile, flexural properties Thermogravimetric Differential scanning calorimetry	[67]
Polyolefin	Coir coconut fibers	Injection molding	Mechanical Viscoelastic Thermal properties	[68]
LDPE	Pineapple fibers	Compression molding	Flexural and Izod impact Water absorption Thermal resistance	[69]
LDPE	Jute fibers	Compression molding	Mechanical properties Moisture intake	[70]
HDPE	Jute fibers	Sandwich technique	Fracture morphology Mechanical performance	[71]
HDPE	Oil-palm	Injection molding	Thermal properties Dynamic mechanical properties	[72]
PET	Moringa Olefin Fruit fibers	Hot press	Thermal Mechanical Morphological	[73]
PET	Hemp	Vacuum infusion	Flexural Tensile Water absorption	[74]
HIPS	Sisal	Compression molding	Tensile Fracture	[75]
Polycarbonate (PC)	Carbon fiber	Injection molding	Thermomechanical Tensile Flexural Izod	[76]
Polypropylene	Kenaf Pineapple	Hot press molding compression	Tensile Flexural Water absorption Surface morphology	[77]

(continued)

Table 2.9 (continued)

Matrix	Reinforcement	Fabrication process	Properties observed	References
Polypropylene	Flax	Hot press technique	Water absorption Sound transmission loss test Tensile Microscopic analysis	[78]
Polypropylene	Coir fibers	Melt-blending	Friction and wear	[79]
Polypropylene	Banana Glass	Twin-screw extruder	Tensile Flexural Impact Water absorption	[80]
PVC	Wood flour Zinc borate	Twin screw extruder + melt press	Thermal properties Mechanical properties	[81]
Polypropylene	Marble waste	Injection molding	Moisture absorption Surface morphology Mechanical properties	[82]

Table 2.10 Merits and demerits of thermoset resins

Type of resin	Advantages	Disadvantages
Epoxy resin	High water resistance High thermal and mechanical properties Low curing time Durability	Difficult to process Corrosive amine hardener More expensive than vinyl ester
Vinyl ester	High chemical resistance High mechanical properties than polyester	High curing shrinkage High styrene content Needs post curing More expensive
Polyester	Lowest cost Easy to use	Limited access Moderate mechanical properties Huge curing shrinkage High gas emissions
Phenolic	High fire retardance	Hard to develop

mechanical properties. Asim et al. investigated the dynamic and mechanical characteristics of kenaf and pineapple fiber-reinforced phenolic hybrid composites [91]. Test samples were prepared using a hydraulic hot press, with fibers ground to 0.1–0.8 mm, and compressed for 8 min to obtain the desired shape for testing purposes. The outcomes demonstrated that kenaf composites exhibited superior mechanical properties, whereas pineapple composites revealed minimal stability under phenolic resin, although the difference was not substantially comparative [92]. The wear resistance of natural corn stalk fiber-reinforced phenolic resin composites was enhanced

Table 2.11 Mechanical properties of thermoset resins

Polymer	Density (g/cm^3)	Tensile strength (MPa)	Young's modulus (GPa)	Elongation (%)	Glass transition Temp. (°C)
Epoxy resin (EP)	1.2–1.3	600.0	80.0	1.3	70.0–167
Urea formaldehyde (UF)	1.5–1.6	65.0	9.0	0.8	130.0
Melamine formaldehyde (MF)	1.5–1.6	65.0	12.0	0.6	150.0
Phenol formaldehyde (PF)	1.2	45.0	6.5	1.2	170.0
Unsaturated polyester (UPE)	1.1	60.0	3.4	2.0	110.0–188.0
Rigid thermoset polyurethane (RPU)	1.2	60.0	2.2	90.0	– 150.0–0.0

using the wet granulation technique [93]. The influence of phenolic resin on brake friction materials was evaluated by varying the percentage of resin matrix in the composite weight, and results indicated that the wet granulation method improved the surface morphology of friction materials and enhanced their tribological behavior.

Singha et al. conducted research on the synthesis and characterization of pine needle-reinforced resorcinol–formaldehyde (RF) biocomposites [94]. Composite sheets with varying fiber loadings were fabricated using the compression molding technique to assess their mechanical properties. The results indicated that the mechanical properties of pine needle composites were significantly enhanced compared to neat RF resin composites. The composite strength increased with fiber loadings, suggesting the potential for these materials to replace synthetic fibers. Singha et al. also explored the mechanical, thermal, and morphological properties of Grewia optivar fiber-reinforced resorcinol–formaldehyde polymer composites [95]. These composites were fabricated using the compression molding technique, with fibers utilized in particulate form for fabrication. Test specimens were prepared by varying the fiber loadings by weight for material properties testing. The findings demonstrated an increase in mechanical properties up to 30% fiber loadings, followed by a slight decline due to weak bonding between the matrix and reinforcement material. Thakur et al. investigated the mechanical properties of hibiscus sabdariffa fiber-reinforced urea-formaldehyde (UF) green composites [96]. Their results indicated an improvement in mechanical properties, with a substantial difference observed between fiber composites and neat UF resin composites. Fiber composites exhibited significantly greater strength than pure resin composites. Singha et al. also focused on the fabrication and characterization of saccharum ciliate fiber-reinforced resorcinol

Table 2.12 Recent research on thermoset hybrid composites and fabrication techniques

Matrix	Reinforcement	Fabrication process	Properties observed	References
Epoxy	Hemp fibers Flax fibers	Hand layup	Mechanical Moisture analysis Contact angle measurement	[98]
Epoxy	Jute fibers Glass fibers	Vacuum infusion	Compressive strength Shear strength	[90]
Epoxy	Crab shell Sisal fibers	Compression molding	Mechanical Chemical Morphological	[99]
Epoxy	Woven cotton Bamboo	Compression molding	Flexural Tensile Impact	[100]
Epoxy	Basalt fiber UHMWPE	Compression molding	Impact Tensile Compression	[101]
Epoxy	Glass Basalt woven	Resin transfer molding technique	Mechanical properties Impact behavior	[102]
Epoxy	Pineapple Basalt	Compression molding	Mechanical properties Tensile Flexural Damping	[103]
Epoxy	Sisal Banana Coir	Compression molding	Hardness Tensile Flexural Water absorption	[104]
Epoxy	Kevlar Cocos Nucifera	Compression molding + hand layup	Thermal Mechanical	[105]
Epoxy	Carbon Flax	Hand layup	Impact strength	[106]
Epoxy	Jute Basalt	Hand layup	Mechanical properties Water absorption	[107]
Epoxy	Bamboo Kenaf	Hand layup	Thermal properties	[108]
Epoxy	Banana Flax	Hand layup	Tensile strength Flexural strength Thermal properties	[109]
Epoxy	Neem Banyan leaf	Hand layup	Thermal stability Thermogravimetry	[110]
Epoxy	Coir Kevlar	Hand layup	Mechanical properties	[111]

(continued)

Table 2.12 (continued)

Matrix	Reinforcement	Fabrication process	Properties observed	References
Vinyl ester	Glass	The Electroless plating method	Thermal wear analysis Mechanical properties	[112]
Vinyl ester	Flax Basalt	Resin infusion and hand layup	Impact strength Mechanical characteristics	[113]
Vinyl ester	Flax Basalt	Vacuum assisted resin infusion	Interlaminar strength Fracture toughness	[114]
Vinyl ester	Flax Basalt	Vacuum infusion	Fracture toughness Water absorption	[115]
Epoxy	Napier Glass	Cold Pressing	Tensile properties	[116]
Epoxy resin	Sisal Coir	Cold Pressing	Moisture absorption Fire retardation	[117]
Epoxy resin	Long kenaf Woven glass	Cold Pressing	Fracture toughness Water absorption	[118]
PLA	Sisal Hemp fibers	Injection molding + Extrusion	Mechanical properties Thermal stability	[119]

formaldehyde green composites [97]. Samples were manufactured using a compression molding machine, and the mechanical properties of these test specimens were evaluated. The results suggested that the saccharum ciliate fibers composites exhibited superior properties compared to matrix composites, indicating the potential for their application in composite fabrication.

Bio-based Resins. Bio-based polymeric resins are synthesized from renewable sources. These resins are systematically categorized into three distinct types according to their physical properties. The first category comprises fully bio-based and biodegradable resins, which include polyhydroxyalkanoates (PHA) and starch. The second category encompasses partially bio-based resins, such as polylactic acid (PLA) and cellulose. The final category consists of partially non-biodegradable resins, including polyethylene terephthalate (PET) and polyethylene. The majority of these bio-based resins are derived from plant materials and plant-based sugars. A detailed examination of the advantages and disadvantages of bio-based resins is provided in Table 2.13.

Table 2.13 Merits and demerits of bio-based resins

Resin	Merits	Demerits
Cellulose	Inexpensive Easily available Ease to adjust Intermediate impact resistance Intermediate heat resistance Eco-friendly	Relative low decomposition High moisture absorption
Polyhydroxy alkenoates (PHA)	Biodegradable High molecular weight	Low stability Fragile Low decomposition costly
Polylactic acid	Nontoxic nature High strength High modulus Low cost	Brittle Poor impact strength Low thermal degradation
Starch	Low cost Biodegradable	Hard to process Water sensitive Brittle

2.6 Integration of Nanoparticles in Natural Fiber Hybrid Composites

Hybrid composites are not free from defects due to fabrication errors, voids within the composite, weak bonding between the matrix and reinforcement, environmental influences, errors during the development and curing phases, scratches on the fiber materials, impurities, and an unbalanced proportion of matrix and reinforcement, all of which adversely affect the overall performance of the composite materials [120]. Mitigating the causes of component degradation can help maintain the efficacy of composite materials. The incorporation of new constituents and the advancement of quality control techniques are essential to enhancing the performance of hybrid composites. One effective approach to address these issues is the integration of nanoparticles or nano constituents. Recent studies indicate that the inclusion of nanofillers in composite materials results in significant improvements in mechanical properties such as tensile strength, flexural strength, creep strength, tensile strain, shear modulus, impact strength, and thermal properties as compared to individual polymer-based hybrid composites [121]. Currently, researchers employ three types of nanofillers to enhance the performance of composite materials. These are classified into three main groups: nanotubes, nanoparticles, and nanolayers. Nanofillers are utilized based on their particle sizes—one-dimensional, two-dimensional, and three-dimensional. Nanomaterials are larger than macro-level particles, individual atoms, and possess a substantial surface area per unit volume. Most of these materials are derived from metals, carbon, polymers, and metal oxides. Examples of widely used metal-based nanofillers and nanomaterials include nano clay, carbon nanotubes, carbon black, and graphene.

Pappu et al. conducted research on underutilized materials, such as fly ash and sisal fibers, for the development of micro and nanocomposites intended for multifunctional applications [122]. The hand layup technique, in conjunction with compression molding, was employed in the fabrication of the composites, where fly ash and sisal fibers served as reinforcement while epoxy acted as the matrix material. A total of two composites were fabricated for comparative analysis. Results suggested that the incorporation of natural fibers and nanofillers enhances the composite's performance, notably due to the cellulose content in sisal fiber and the presence of silica and alumina in fly ash, which contribute to the enhancement of mechanical properties. Majeed et al. investigated food packaging materials made from natural fiber hybrid composites with nanofillers in the form of nano clay [72], and devised a novel method to improve the bonding between the matrix and reinforcement. The integration of nano clay into the hybrid composite demonstrated significant progress in mechanical and thermal properties. However, these materials are not universally applicable for all packaging types due to higher biodegradability and moisture absorption characteristics compared to synthetic fibers. Essabir et al. examined the mechanical and thermal properties of clay and palm fiber hybrid composites [123]. Their findings indicated that the integration of oil palm with clay particles yielded improvements in tensile strength and Young's modulus. Arrakhiz et al. reported analogous tests, documenting the impact of mechanical and thermal properties in pine fiber reinforced with clay particle hybrid composites [124]; their findings were consistent with prior research. The inclusion of clay particles resulted in the mitigation of voids on the composite surface, albeit to the detriment of surface properties and overall composite performance.

Nanoclay particles are extensively utilized as filler materials due to their availability and cost-effectiveness, being classified within the silicate group of nanomaterials. The incorporation of nanoclay particles into composite materials can enhance mechanical, physical, and chemical properties. Moreover, the size and shape of these particles significantly influence the strength and bonding between the reinforcement and matrix material in composites. Toyota automobile industries pioneered the use of nanoclay particles in polymer composite fabrication [125]; in their research, the author highlighted the pivotal role of clay particles as filler material in establishing interfacial bonds between the matrix and reinforcement, alongside improving the physical and fracture toughness of the composite material. Owing to the enhancement of material properties facilitated by nanoclay particles, these materials have found applications in various functional and structural domains. The nanoparticle microstructure within the material is instrumental in the advancement of mechanical properties.

Nanoclays are typically available in various forms, such as nanoclay platelets, calcined nanoclay, and natural montmorillonite (CB). Assaedi et al. conducted an investigation into the influence of nanoclay platelets, also referred to as cloisite 30B, on the thermal and mechanical properties of fly ash geopolymer [126]. In this study, the platelets were incorporated at differing weight percentages, specifically 1%, 2%, and 3% loadings. The findings indicated that the inclusion of nanoclay platelets in the geopolymer resulted in a slight enhancement of mechanical properties. Notably, both flexural strength and flexural modulus exhibited significant increases with rising nanoclay loading up to a 2% weight threshold. Beyond 2% weight loading, a marginal decline in material properties was observed, attributed to porosity and inadequate dispersion. It became evident that the continuous addition of filler materials to the geopolymers does not enhance its mechanical properties. Additionally, Assaedi addressed the conditions regarding temperature variation and examined this aspect through thermo-gravimetric analysis in terms of weight. It was noted that the most significant weight loss occurred with a 3% nanoclay loading, primarily due to increased friction.

Hakamy et al. examined the impact of calcined nanoclay (CNC) on the mechanical and microstructural properties of hemp fiber reinforced Portland cement nanocomposites [127]. Calcined nanoclay was synthesized through a heat treatment process of nanoclay particles at temperatures of 800, 850, and 900 °C for a duration of two hours. It was determined that nanoclay achieved a calcined state at 900 °C. Experimental findings indicate that the mechanical properties, including flexural strength and fracture toughness, were enhanced by the incorporation of CNC into the composite material. Calcined nanoclay particles were introduced at variable weight percentages of 1, 2, and 3%. The optimal hemp fabric content was identified at 6.9%, with CNC optimizing at 1%. Eesaee et al. investigated the effect of nanoclay particles in reinforced glass/woven fiber composites with phenolic resins [128], focusing specifically on natural montmorillonite (CN) and artificial montmorillonite (CB). The filler material enhanced composite performance due to the effective dispersion of nanoclay particles on the composite surface, which facilitated improved interfacial bonding with the matrix and reinforcement. It was observed that the incorporation of 2.5 wt% of the filler material increased the elastic modulus by up to 38% for CN and 43% for CB. Conversely, aging due to exposure to water molecules and hydroxyl groups of the phenolic matrix resulted in a reduction in the composite's strength.

Avil et al. investigated the impact of porosity and failure strength with the inclusion of nanoclay in glass fiber reinforced epoxy composites [129]. The findings revealed that the addition of the filler material resulted in a reduction of porosity by 73.2 at 10 wt%. Binu et al. examined the influence of nanoclay, specifically Cloisite 15A, on the mechanical performance and thermal attributes of polyester-based glass fiber composites [130]. Their study identified the peak mechanical performance at 1% filler material loading. According to their analysis, the glass transition temperature occurred between 95 and 110 °C. John et al. explored the dynamic mechanical and thermal characteristics of cyanate ester syntactic foams incorporating nanoclay material [131]. The research indicated a notable enhancement in mechanical properties such as tensile and flexural strength moduli upon the integration of nanoclay particles. While the incorporation of nanoclay increases the toughness of the composite, thermal properties are minimally affected by the addition of filler materials. Maharsia et al. [132] probed the mechanical attributes of syntactic foams with nanoclay reinforcement. The results indicated that the presence of nanoclay (2 and 5% by weight) improved mechanical properties, although a reduction in modulus was observed due to suboptimal bonding between nanoclay particles and the matrix material. Withers et al. [133] assessed the effect of Cloisite 25A montmorillonite (MMT) nanoclay as a filler material in epoxy-based composites. The nanocomposites were fabricated by varying the weight percentages of nanoparticle inclusion. The findings determined that optimal mechanical properties occur at a 2 wt% MMT clay addition, while enhanced surface properties are observed at lower percentage loadings of MMT clay particles. Mo-lin Chan et al. [134] investigated the reinforcement mechanism in nanoclay polymer composites. Various mechanical tests were performed by adjusting the nanoparticle content within the composite material. Notably, results demonstrated that tensile strength and modulus of the composite improved by 34% and 25%, respectively. The formation of boundaries between the nanoclay clusters and the epoxy resin can serve to refine the matrix grains and further augment the composite's strength. A selection of research publications pertaining to nanofillers in composite materials is presented in Table 2.14.

2.7 Conclusions

In this chapter, fiber-based hybrid composites are evaluated specifically concerning their properties, which include surface morphology, thermal stability, dynamic mechanical behavior, mechanical properties, moisture absorption, fire retardation, and applications. The concept of hybridization gained prominence not only due to the enhanced performance of the resulting products but also because of the potential to overcome the limitations that restrict the applicability of natural fibers in specialized structures. There exists substantial potential for a significant proportion of natural fibers to be integrated into conventional engineered reinforced composite products, representing a considerable advancement from both ecological and economic perspectives. The quality of natural fibers depends on growth conditions, and the growth period impacts the composite's performance. Despite the numerous advantages of utilizing natural fiber in commercial applications, its application is confined due to its biodegradability and flammable nature compared to synthetic fiber composites. To address this issue, researchers have initiated chemical treatment of natural fibers under various conditions. These modifications have enhanced fiber strength and improved the bonding capacity between matrix and reinforcement. Moreover, researchers have shown interest in incorporating nanofillers into composite materials. The inclusion of nanoparticles in hybrid composites has played a crucial role in enhancing their properties. The addition of these nanofiller materials in the hybrid composite reduces surface properties, fracture toughness, void elimination, and water absorption, while increasing dynamic mechanical properties. It is noteworthy that hybrid composite materials are revolutionizing real-world commercial applications, including biomedical, aerospace, shipping industry, automotive, construction, drug delivery, wound dressing, and gas filtration applications. In the future, modifications in fabrication techniques and the incorporation of nanofillers may lead to further enhancements and predictive capabilities of results, thus attracting researchers to explore multiple applications.

Table 2.14 Merits and demerits of bio-based resins

Fibers/Nanofillers	Polymer	Investigations	References
Glass fibers/Silica	Epoxy resin	Elastic properties	[135]
Pineapple/Silica	Epoxy resin	Tensile fatigue Fracture toughness	[136]
Rubber/Silica	Epoxy resin	Vibration and damping	[137]
Glass/Kevlar/Silica	Epoxy resin	Fracture toughness Tensile strength Flexural strength	[138]
Banana/Sisal/Glass/Redmud	Polyester resin	Tensile strength Flexural strength Impact strength	[139]
Silica/Glass/Graphite	Polypropylene	Thermo Mechanical Elastic properties	[140]
Basalt/Clay	Polypropylene	Impact Properties	[141]
Carbon fiber/Carbon nanotube	Polyethylene	Mechanical properties	[142]
Rubber/Carbon nanotube	–	Stress–strain analysis	[143]
Carbon fiber/Carbon nanotube	–	Mechanical properties	[144]
Bamboo/Clay	Polyvinyl alcohol	Thermal, Physical Mechanical properties	[145]
Pineapple/Carbon/Jute	Epoxy resin	Tensile, flexural and fracture toughness	[146]
Sugar palm/Clay	Polyester	Thermal and mechanical properties	[147]
Glass/Clay	Polyester	Mechanical properties	[148]
Glass/Clay	Vinyl ester	Thermal and flame retardancy studies	[149]
Glass/Silica	Vinyl ester	Mechanical properties	[150]
Glass/Clay	Vinyl ester	Mechanical, thermal, vibration properties	[151–153]
Glass/Clay	Vinyl ester	Mechanical properties	[154]
Basalt/graphene oxide	Epoxy resin	Interlaminar shear strength Fracture toughness	[155]

References

1. Atmakuri A, Palevicius A, Vilkauskas A, Janusas G (2020) Review of hybrid fiber based composites with nano particles–material properties and applications. Polym Basel MDPI 12(9):1–30. ISSN 2073-4360. Art. no. 2088. https://doi.org/10.3390/polym12092088
2. Wambua P, Ivens J, Verpoest I (2003) Natural fibers: can they replace glass in fiber reinforced plastics. Compos Sci Technol 63:1259–1264
3. Rajak DK, Pagar DD, Menezes PL, Linul E (2019) Fiber-reinforced polymer composites: manufacturing, properties, and applications. Polymers 11:1667
4. Saba N, Paridah MT, Abdan K, Ibrahim NA (2016) Dynamic mechanical properties of oil palm nano filler/kenaf/epoxy hybrid nanocomposites. Constr Build Mater 124:133–138

5. Ahmad F, Choi HS, Park MK (2015) A review: Natural fiber composites selection in view of mechanical, light weight and economic properties. Macromol Mater Eng 300:10–24
6. Kiruthika AV (2017) A review on physico-mechanical properties of bast fibre reinforced polymer composites. J Build Eng 9:91–99
7. George J, Sreekala MS, Thomas S (2001) A review on interface modification and characterization of natural fiber reinforced plastic composites. Polym Eng Sci 41:1471–1485
8. Faruk O, Bledzki AK, Fink HP, Sain M (2012) Biocomposites reinforced with natural fibers: 2000–2010. Prog Polym Sci 37:1552–1596
9. Biswas B, Kandola BK (2011) The effect of chemically reactive type flame retardant additives on flammability of PES toughened epoxy resin and carbon fiber-reinforced composites. Polym Adv Technol 22:1192–1204
10. John MJ, Anandjiwala RD (2008) Recent developments in chemical modification and characterization of natural fiber-reinforced composites. Polym Compos 29:187–207
11. Galos J (2020) Thin-ply composite laminates: a review. Compos Struct 236:1–12
12. Grefe H, Kandula MW, Dilger K (2020) Influence of the fibre orientation on the lap shear strength and fracture behavior of adhesively bonded composite metal joints at high strain rates. Int J Adhes Adhes 97:1–7
13. Ou Y, Gattas JM, Fernando D, Torero JL (2020) Experimental investigation of a timber-concrete floor panel system with a hybrid glass fibre reinforced polymer-timber corrugated core. Eng Struct 203:1–15
14. Mishra S, Mohanty AK, Drzal LT, Misra M, Parija S, Nayak SK, Tripathy SS (2003) Studies on mechanical performance of biofibre/glass reinforced polyester hybrid composites. Compos Sci Technol 63:1377–1385
15. Jawaid M, Abdul Khalil HPS (2011) Cellulosic/synthetic fibre reinforced polymer hybrid composites: a review. Carbohydr Polym 86:1–18
16. John MJ, Thomas S (2008) Biofibres and biocomposites. Carbohydr Polym 71:343–364
17. Bhoopathi R, Ramesh M, Deepa C (2014) Fabrication and property evaluation of banana-hemp-glass fiber reinforced composites. Procedia Eng 97:2032–2041
18. Alavudeen A, Rajini N, Karthikeyan S, Thiruchitrambalam M, Venkateshwaren N (2015) Mechanical properties of banana/kenaf fiber-reinforced hybrid polyester composites: effect of woven fabric and random orientation. Mater Des 66:246–257
19. Ramesh M, Palanikumar K, Reddy KH (2013) Mechanical property evaluation of sisal-jute-glass fiber reinforced polyester composites. Compos Part B 48:1–9
20. Almeida JHS Jr, Amico SC, Botelho EC, Amado FDR (2013) Hybridization effect on the mechanical properties of curaua/glass fiber composites. Compos Part B 55:492–497
21. Velmurugan R, Manikandan V (2007) Mechanical properties of palmyra/glass fiber hybrid composites. Compos Part A Appl Sci Manuf 38:2216–2226
22. Kumar A, Singh RC, Chaudhary R (2020) Recent progress in production of metal matrix composites by stir casting process—an overview. Mater Today Proc 21:1453–1457
23. Rajak DK, Pagar DD, Kumar R, Pruncu CI (2019) Recent progress of reinforcement materials: a comprehensive overview of composite materials. J Mater Res Technol 8:6354–6374
24. Sharma AK, Bhandari R, Aherwar A, Rimasauskiene R (2020) Matrix materials used in composites: a comprehensive study. Mater Today Proc 21:1559–1562
25. Bharath KN, Basavarajappa S (2016) Applications of biocomposite materials based on natural fibers from renewable resources: a review. Sci Eng Compos Mater 23:123–133
26. Djafari Petroudy SR (2017) Physical and mechanical properties of natural fibers. Adv High Strength Nat Fibre Compos Constr 03:59–83
27. Mohanty AK, Misra M, Hinrichsen G (2000) Biofibres, biodegradable polymers and biocomposites: an overview. Macromol Mater Eng 276:1–24
28. Gholampour A, Ozbakkaloglu T (2020) A review of natural fiber composites: properties, modification and processing techniques, characterization, applications. J Mater Sci 55:829–892
29. Wang X, Xie J, Zhang H, Zhang W, An S, Chen S, Luo C (2016) Determining the lignin distribution in plant fiber cell walls based on chemical and biological methods. Cellulose 26:4241–4252

30. Sood M, Dwivedi G (2018) Effect of fiber treatment on flexural properties of natural fiber reinforced composites: a review. Egypt J Pet 27:775–783
31. Li X, Tabil LG, Panigrahi S (2007) Chemical treatments of natural fiber for use in natural fiber-reinforced composites: a review. J Polym Environ 15:25–33
32. Mohanty AK, Misra M, Drzal LT (2001) Surface modifications of natural fibers and performance of the resulting biocomposites: an overview. Compos Interfaces 8:313–343
33. Ray D, Sarkar BK, Rana AK, Bose NR (2001) Effect of alkali treated jute fibres on composite properties. Bull Mater Sci 24:129–135
34. Mishra S, Misra M, Tripathy SS, Nayak SK, Mohanty AK (2001) Graft copolymerization of acrylonitrile on chemically modified sisal fibers. Macromol Mater Eng 286:107–113
35. Ali A, Shaker K, Nawab Y, Jabbar M, Hussain T, Militky J, Baheti V (2018) Hydrophobic treatment of natural fibers and their composites—a review. J Ind Text 47:2153–2183
36. Rong MZ, Zhang MQ, Liu Y, Yang GC, Zeng HM (2001) The effect of fiber treatment on the mechanical properties of unidirectional sisal-reinforced epoxy composites. Compos Sci Technol 61:1437–1447
37. Kumar R, Obrai S, Sharma A (2011) Chemical modifications of natural fiber for composite material. Pelagia Res Libr 2:219–228
38. Valadez-Gonzalez A, Cervantes-Uc JM, Olayo R, Herrera-Franco PJ (1999) Effect of fiber surface treatment on the fiber-matrix bond strength of natural fiber reinforced composites. Compos Part B Eng 30:309–320
39. Joseph PV, Mathew G, Joseph K, Groeninckx G, Thomas S (2003) Dynamic mechanical properties of short sisal fibre reinforced polypropylene composites. Compos Part A Appl Sci Manuf 34:275–290
40. Idicula M, Malhotra SK, Joseph K, Thomas S (2005) Dynamic mechanical analysis of randomly oriented intimately mixed short banana/sisal hybrid fibre reinforced polyester composites. Compos Sci Technol 65:1077–1087
41. Paul S, Nanda P, Gupta R (2003) PhCOCl-Py/basic alumina as a versatile reagent for benzoylation in solvent-free conditions. Molecules 8:374–380
42. Pinem JA, Wardani AK, Aryanti PTP, Khoiruddin K, Wenten IG (2019) Hydrophilic modification of polymeric membrane using graft polymerization method: a mini review. IOP Conf Ser Mater Sci Eng 547:1–12
43. Park JM, Quang ST, Hwang BS, DeVries KL (2006) Interfacial evaluation of modified Jute and Hemp fibers/polypropylene (PP)-maleic anhydride polypropylene copolymers (PP-MAPP) composites using micromechanical technique and nondestructive acoustic emission. Compos Sci Technol 66:2686–2699
44. Mohanty S, Nayak SK, Verma SK, Tripathy SS (2004) Effect of MAPP as a coupling agent on the performance of Jute-PP composites. J Reinf Plast Compos 23:625–637
45. Sreekala MS, Kumaran MG, Joseph S, Jacob M, Thomas S (2000) Oil palm fibre reinforced phenol formaldehyde composites: Influence of fibre surface modifications on the mechanical performance. Appl Compos Mater 7:295–329
46. Suwanruji P, Tuechart T, Smitthipong W, Chollakup R (2017) Modification of pineapple leaf fiber surfaces with silane and isocyanate for reinforcing thermoplastic. J Thermoplast Compos Mater 30:1344–1360
47. Shih YF, Chang WC, Liu WC, Lee CC, Kuan CS, Yu YH (2014) The inclusion of PALF made the thermoplastic matrix stiffer, especially for the LDPE composites. J Taiwan Inst Chem Eng 45:2039–2046
48. Sanjay MR, Madhu P, Jawaid M, Senthamaraikannan P, Senthil S, Pradeep S (2018) Characterization and properties of natural fiber polymer composites: a comprehensive review. J Clean Prod 172:566–581
49. Fei ME, Xie T, Liu W, Chen H, Qiu R (2017) Surface grafting of bamboo fibers with 1,2-epoxy-4-vinylcyclohexane for reinforcing unsaturated polyester. Cellulose 24:5505–5514
50. Seki Y (2009) Innovative multifunctional siloxane treatment of jute fiber surface and its effect on the mechanical properties of jute/thermoset composites. Mater Sci Eng A 508:247–252

51. Jarukumjorn K, Suppakarn N (2009) Effect of glass fiber hybridization on properties of sisal fiber-polypropylene composites. Compos Part B Eng 40:623–627
52. Schmidt TM, Goss TM, Amico SC, Lekakou C (2009) Permeability of hybrid reinforcements and mechanical properties of their composites molded by resin transfer molding. J Reinf Plast Compos 38:2839–2850
53. Nayak SK, Mohanty S, Samal SK (2009) Influence of short bamboo/glass fiber on the thermal, dynamic mechanical and rheological properties of polypropylene hybrid composites. Mater Sci Eng A 523:32–38
54. Vega-Hernández MA, Rosas-Aburto A, Vivaldo-Lima E, Vázquez-Torres H, Cano-Díaz GS, Pérez-Salinas P, Hernández-Luna MG, Alcaraz-Cienfuegos J, Zolotukhin MG (2019) Development of polystyrene composites based on blue agave bagasse by in situ RAFT polymerization. J Appl Polym Sci 136:1–14
55. Haneefa A, Bindu P, Aravind I, Thomas S (2008) Studies on tensile and flexural properties of short banana/glass hybrid fiber reinforced polystyrene composites. J Compos Mater 42:1471–1489
56. Hao X, Zhou H, Mu B, Chen L, Guo Q, Yi X, Sun L, Wang Q, Ou R (2020) Effects of fiber geometry and orientation distribution on the anisotropy of mechanical properties, creep behavior and thermal expansion of natural fiber/HDPE composites. Compos Part B Eng 185:1–9
57. Robledo-Ortíz JR, González-López ME, Rodrigue D, Gutiérrez-Ruiz JF, Prezas-Lara F, Pérez-Fonseca AA (2020) Improving the compatibility and mechanical properties of natural fibers/green polyethylene biocomposites produced by rotational molding. J Polym Environ 28:1040–1049
58. Paran SMR, Naderi G, Shokoohi S, Ebadati J, Dubois C (2019) Mechanical and thermal properties of green thermoplastic elastomer vulcanizate nanocomposites based on poly (vinyl chloride) and nitrile butadiene rubber containing organoclay and rice straw natural fibers. J Polym Environ 27:2017–2026
59. Ji B, Gao H (2004) Mechanical properties of nanostructure of biological materials. J Mech Phys Solids 52:1963–1990
60. Zare Y, Garmabi H (2015) A developed model to assume the interphase properties in a ternary polymer nanocomposite reinforced with two nanofillers. Compos Part B Eng 75:29–35
61. Boussehel H, Mazouzi DE, Belghar N, Guerira B, Lachi M (2019) Effect of chemicals treatments on the morphological, mechanical, thermal and water uptake properties of polyvinyl chloride/palm fibers composites. Rev Compos Mater Av 29:1–8
62. Ihamouchen C, Djidjelli H, Boukerrou A, Krim S, Kaci M, Martinez JJ (2012) Effect of surface treatment on the physicomechanical and thermal properties of high-density polyethylene/olive husk flour composites. J Appl Polym Sci 123:1310–1319
63. Kaci M, Djidjelli H, Boukerrou A, Zaidi L (2007) Effect of wood filler treatment and EBAGMA compatibilizer on morphology and mechanical properties of low-density polyethylene/olive husk flour composites. Express Polym Lett 1:467–473
64. Haseena AP, Dasan KP, Namitha R, Unnikrishna G, Thomas SA (2004) Investigation on interfacial adhesion of short sisal/coir hybrid fibre reinforced natural rubber composites by restricted equilibrium swelling technique. Compos Interfaces 11:489–513
65. Prasantha Kumar R, Thomas S (2001) Interfacial adhesion in sisal fiber/SBR composites: an investigation by the restricted equilibrium swelling technique. J Adhes Sci Technol 15:633–652
66. Sari PS, Thomas S, Spatenka P, Ghanam Z, Jenikova Z (2019) Effect of plasma modification of polyethylene on natural fibre composites prepared via rotational moulding. Compos Part B Eng 177:1–10
67. Boran Torun S, Pesman E, Donmez Cavdar A (2019) Effect of alk ali treatment on composites made from recycled polyethylene and chestnut cupula. Polym Compos 40:4442–4451
68. Salazar MAH, Aguirre JPC, Navarro SG, Blay LR (2020) Injection molding of coir coconut fiber reinforced polyolefin blends: mechanical, viscoelastic, thermal behavior and three-dimensional microscopy study. Polymers 12:1507

69. Rahman H, Alimuzzaman S, Sayeed MMA, Khan RA (2019) Effect of gamma radiation on mechanical properties of pineapple leaf fiber (PALF)-reinforced low-density polyethylene (LDPE) composites. Int J Plast Technol 23:229–238
70. Sahadat Hossain M, Razzak M, Uddin MB, Chowdhury AMS, Khan RA (2019) Physico-mechanical properties of jute fiber-reinforced LDPE-based composite: effect of disaccharide (sucrose) and gamma radiation. Radiat Eff Defects Solids 175:516–528
71. Sayem ASM, Haider J, Sayeed MA (2019) Development and characterisation of multi-layered jute fabric-reinforced HDPE composites. J Compos Mater 54:1831–1845
72. Essabir H, Boujmal R, Bensalah MO, Rodrigue D, Bouhfid R, Qaiss AEK (2016) Mechanical and thermal properties of hybrid composites: oil-palm fiber/clay reinforced high density polyethylene. Mech Mater 98:36–43
73. Nayak S, Kumar Khuntia S (2019) Development and study of properties of Moringa oleifera fruit fibers/polyethylene terephthalate composites for packaging applications. Compos Commun 15:113–119
74. Ahmad MAA, Abdul Majid MS, Ridzuan MJM, Mazlee MN, Gibson AG (2018) Dynamic mechanical analysis and effects of moisture on mechanical properties of interwoven hemp/polyethylene terephthalate (PET) hybrid composites. Constr Build Mater 179:265–276
75. Antich P, Vázquez A, Mondragon I, Bernal C (2006) Mechanical behavior of high impact polystyrene reinforced with short sisal fibers. Compos Part A Appl Sci Manuf 37:139–150
76. Andrzejewski J, Misra M, Mohanty AK (2018) Polycarbonate biocomposites reinforced with a hybrid filler system of recycled carbon fiber and biocarbon: preparation and thermomechanical characterization. J Appl Polym Sci 135:1–14
77. Feng NL, Malingam SD, Ping CW, Razali N (2019) Mechanical properties and water absorption of kenaf/pineapple leaf fiber-reinforced polypropylene hybrid composites. Polym Compos 41:1255–1264
78. Haris A, Kureemun U, Tran LQN, Lee HP (2019) Water uptake and its effects on mechanical and acoustic properties of flax/polypropylene composite. J. Nat. Fibers 16:1–15
79. Liu L, Wang Z, Yu Y, Fu S, Nie Y, Wang H, Song P (2019) Engineering interfaces toward high-performance Polypropylene/Coir fiber biocomposites with enhanced friction and wear behavior. ACS Sustain. Chem. Eng. 7:18453–18462
80. Samal SK, Mohanty S, Nayak SK (2009) Banana/glass fiber-reinforced polypropylene hybrid composites: fabrication and performance evaluation. Polym—Plast Technol Eng 48:397–414
81. Fang Y, Wang Q, Guo C, Song Y, Cooper PA (2013) Effect of zinc borate and wood flour on thermal degradation and fire retardancy of Polyvinyl chloride (PVC) composites. J Anal Appl Pyrolysis 100:230–236
82. Bakshi P, Pappu A, Patidar R, Gupta MK, Thakur VK (2020) Transforming marble waste into high-performance, water resistant and thermally insulative hybrid polymer composites for environmental sustainability. Polymers 12:1781
83. Bakare FO, Skrifvars M, Åkesson D, Wang Y, Afshar SJ, Esmaeili N (2014) Synthesis and characterization of bio-based thermosetting resins from lactic acid and glycerol. J Appl Polym Sci 131:1–9
84. Ail Mutar M, Attab ZHA (2017) Synthesis, characterization and properties of new unsaturated polyesters resins reinforced with some fillers (Carbon Nano, Nano TiO_2, TiO_2 and ZnO) for composite application. Int J Appl Nat Sci (IJANS) 6:79–100
85. Mohammad NA (2017) Synthesis, characterization and properties of the new unsaturated polyester resins for composite application. University sains malaysia, Malaysia, Masters
86. Yorseng K, Rangappa SM, Pulikkalparambil H, Siengchin S, Parameswaranpillai J (2020) Accelerated weathering studies of kenaf/sisal fiber fabric reinforced fully biobased hybrid bioepoxy composites for semi-structural applications: Morphology, thermo-mechanical, water absorption behavior and surface hydrophobicity. Constr Build Mater 235:1–14
87. Manalo AC, Aravinthan T, Karunasena W, Islam MM (2010) Flexural behaviour of structural fibre composite sandwich beams in flatwise and edgewise positions. Compos Struct 92:984–995

88. Rwawiire S, Tomkova B, Militky J, Jabbar A, Kale BM (2015) Development of a biocomposite based on green epoxy polymer and natural cellulose fabric (bark cloth) for automotive instrument panel applications. Compos Part B Eng 81:149–157
89. Pandey JK, Ahn SH, Lee CS, Mohanty AK, Misra M (2010) Recent advances in the application of natural fiber-based composites. Macromol Mater Eng 295:975–989
90. Alves JLC, Prado KS, de Paiva JMF (2019) Compressive and Interlaminar shear strength properties of biaxial fibreglass laminates hybridized with jute fibre produced by vacuum infusion. J Nat Fibers 16:1–16
91. Asim M, Jawaid M, Paridah MT, Saba N, Nasir M, Shahroze RM (2019) Dynamic and thermo-mechanical properties of hybridized kenaf/PALF reinforced phenolic composites. Polym Compos 40:3814–3822
92. Asim M, Paridah MT, Saba N, Jawaid M, Alothman OY, Nasir M, Almutairi Z (2018) Thermal, physical properties and flammability of silane treated kenaf/pineapple leaf fibres phenolic hybrid composites. Compos Struct 202:1330–1338
93. Liu Y, Wang L, Liu D, Ma Y, Tian Y, Tong J, Senthamaraikannan P, Saravanakumar S (2019) Evaluation of wear resistance of corn stalk fiber reinforced brake friction materials prepared by wet granulation. Wear 432–433:102918
94. Singha AS, Thakur VK (2008) Synthesis and characterization of pine needles reinforced RF matrix based biocomposites. E-J Chem 5:1055–1062
95. Singha AS, Thakur VK (2009) Mechanical, thermal and morphological properties of grewia optiva fiber polymer matrix composites. Polym-Plast Technol Eng 48:201–208
96. Singha AS, Thakur VK (2009) fabrication and characterization of *H. sabdariffa* fiber reinforced green polymer composites. Polym-Plast Technol Eng 48:482–487
97. Singha AS, Thakur VK (2009) Fabrication and characterization of *S. cilliare* fiber reinforced polymer composites. Bull Mater Sci 32:49–58
98. Atmakuri A, Palevicius A, Griskevicius P, Janušas G (2019) Investigation of mechanical properties of hemp and flax fibers hybrid composites for biomedical applications. Mechanika 25:149–155
99. Soundhar A, Kandasamy J (2019) Mechanical, chemical and morphological analysis of crab shell/sisal natural fiber hybrid composites. J Nat Fibers 16:1–15
100. Aruchamy K, Pavayee Subramani S, Palaniappan SK, Sethuraman B, Velu Kaliyannan G (2020) Study on mechanical characteristics of woven cotton/bamboo hybrid reinforced composite laminates. J Mater Res Technol 9:718–726
101. Yang Z, Liu J, Wang F, Li S, Feng X (2019) Effect of fiber hybridization on mechanical performances and impact behaviors of basalt fiber/UHMWPE fiber reinforced epoxy composites. Compos Struct 229:1–13
102. Sarasini F, Tirillò J, Valente M, Valente T, Cioffi S, Iannace S, Sorrentino L (2013) Effect of basalt fiber hybridization on the impact behavior under low impact velocity of glass/basalt woven fabric/epoxy resin composites. Compos Part A Appl Sci Manuf 47:109–123
103. Doddi PRV, Chanamala R, Dora SP (2019) Dynamic mechanical properties of epoxy based PALF/basalt hybrid composite laminates. Mater Res Express 6:1–8
104. Balaji A, Sivaramakrishnan K, Karthikeyan B, Purushothaman R, Swaminathan J, Kannan S, Udhayasankar R, Madieen AH (2019) Study on mechanical and morphological properties of sisal/banana/coir fiber-reinforced hybrid polymer composites. J Braz Soc Mech Sci Eng 41:1–10
105. Naveen J, Jawaid M, Zainudin ES, Sultan MTH, Yahaya R (2019) Mechanical and moisture diffusion behavior of hybrid Kevlar/Cocos nucifera sheath reinforced epoxy composites. J Mater Res Technol 8:1308–1318
106. Al-Hajaj Z, Sy BL, Bougherara H, Zdero R (2019) Impact properties of a new hybrid composite material made from woven carbon fibers plus flax fibers in an epoxy matrix. Compos Struct 208:346–356
107. Prasad L, Saini A, Kumar V (2019) Mechanical performance of jute and basalt fiber geo-grid-reinforced epoxy hybrid composite material. J Nat Fibers 16:1–11

108. Chee SS, Jawaid M, Sultan MTH, Alothman OY, Abdullah LC (2019) Evaluation of the hybridization effect on the thermal and thermo-oxidative stability of bamboo/kenaf/epoxy hybrid composites. J Therm Anal Calorim 137:55–63
109. Srinivasan VS, Rajendra Boopathy S, Sangeetha D, Vijaya Ramnath B (2014) Evaluation of mechanical and thermal properties of banana-flax based natural fibre composite. Mater Des 60:620–627
110. Raja T, Anand P (2019) Evaluation of thermal stability and thermal properties of neem/banyan reinforced hybrid polymer composite. Mater Perform Charact 8:1–10
111. Rashid AHA, Ahmad R, Jaafar M, Roslan MN, Ariffin S (2011) Mechanical properties evaluation of woven coir and kevlar reinforced epoxy composites. Adv Mater Res 277:36–42
112. Anand G, Alagumurthi N, Elansezhian R, Venkateshwaran N (2018) Dynamic mechanical, thermal and wear analysis of Ni-P coated glass fiber/Al_2O_3 nanowire reinforced vinyl ester composite. Alex Eng J 57:621–631
113. Fragassa C, Pavlovic A, Santulli C (2018) Mechanical and impact characterisation of flax and basalt fibre vinylester composites and their hybrids. Compos Part B Eng 137:247–259
114. Almansour FA, Dhakal HN, Zhang ZY (2018) Investigation into mode II interlaminar fracture toughness characteristics of flax/basalt reinforced vinyl ester hybrid composites. Compos Sci Technol 154:117–127
115. Almansour FA, Dhakal HN, Zhang ZY (2017) Effect of water absorption on mode I interlaminar fracture toughness of flax/basalt reinforced vinyl ester hybrid composites. Compos Struct 168:813–825
116. Fatinah TS, Abdul Majid MS, Ridzuan MJM, Hong TW, Amin NAM, Afendi M (2017) Tensile properties of compressed moulded Napier/glass fibre reinforced epoxy composites. J Phys Conf Ser 908:1–9
117. Akash, Girisha KG,Venkatesha Gupta NS, Sreenivas Rao KV (2017) A study on flammability and moisture absorption behavior of sisal/coir fiber reinforced hybrid composites. IOP Conf Ser Mater Sci Eng 191:1–5
118. Salleh Z, Taib YM, Hyie KM, Mihat M, Berhan MN, Ghani MAA (2012) Fracture toughness investigation on long kenaf/woven glass hybrid composite due to water absorption effect. Procedia Eng 41:1667–1673
119. Pappu A, Pickering KL, Thakur VK (2019) Manufacturing and characterization of sustainable hybrid composites using sisal and hemp fibres as reinforcement of poly (lactic acid) via injection moulding. Ind Crops Prod 137:260–269
120. Rosamah E, Hossain MS, Abdul Khalil HPS, Wan Nadirah WO Dungani R, Nur Amiranajwa AS, Suraya NLM, Fizree HM, Mohd Omar AK (2017) Properties enhancement using oil palm shell nanoparticles of fibers reinforced polyester hybrid composites. Adv Compos Mater 26:259–272
121. Boccaccini AR, Erol M, Stark WJ, Mohn D, Hong Z, Mano JF (2010) Polymer/bioactive glass nanocomposites for biomedical applications: a review. Compos Sci Technol 70:1764–1776
122. Pappu A, Thakur VK (2017) Towards sustainable micro and nano composites from fly ash and natural fibers for multifunctional applications. Vacuum 146:375–385
123. Majeed K, Jawaid M, Hassan AABAA, Bakar AA, Khalil HA, Salema AA, Inuwa I (2013) Potential materials for food packaging from nanoclay/natural fibres filled hybrid composites. Mater Des 46:391–410
124. Arrakhiz FZ, Benmoussa K, Bouhfid R, Qaiss A (2013) Pine cone fiber/clay hybrid composite: mechanical and thermal properties. Mater Des 50:376–381
125. El-Sheikhy R, Al-Shamrani M (2017) Interfacial bond assessment of clay-polyolefin nanocomposites CPNC on view of mechanical and fracture properties. Adv Powder Technol 28:983–992
126. Assaedi H, Shaikh FUA, Low IM (2016) Effect of nano-clay on mechanical and thermal properties of geopolymer. J Asian Ceram Soc 190:1–10
127. Hakamy A, Shaikh FUA, Low IM (2015) Effect of calcined nanoclay on microstructural and mechanical properties of chemically treated hemp fabric-reinforced cement nanocomposites. Constr Build Mater 33:395–399

128. Eesaee M, Shojaei A (2014) Effect of nanoclays on the mechanical properties and durability of novolac phenolic resin/woven glass fiber composite at various chemical environments. Compos Part A Appl Sci Manuf 63:149–158
129. Ávila AF, Morais DTS (2009) Modeling nanoclay effects into laminates failure strength and porosity. Compos Struct 87:55–62
130. Binu PP, George KE, Vinodkumar MN (2016) Effect of nanoclay, Cloisite15A on the mechanical properties and thermal behavior of glass fiber reinforced polyester. Procedia Technol 25:843–853
131. John B, Nair CPR, Ninan KN (2010) Effect of nanoclay on the mechanical, dynamic mechanical and thermal properties of cyanate ester syntactic foams. Mater Sci Eng A 527:5435–5443
132. Maharsia RR, Jerro HD (2007) Enhancing tensile strength and toughness in syntactic foams through nanoclay reinforcement. Mater Sci Eng A 454:416–422
133. Withers GJ, Yu Y, Khabashesku VN, Cercone L, Hadjiev VG, Souza JM, Davis DC (2015) Improved mechanical properties of an epoxy glass-fiber composite reinforced with surface organomodified nanoclays. Compos Part B Eng 72:175–182
134. Chan ML, Lau KT, Wong TT, Ho MP, Hui D (2011) Mechanism of reinforcement in a nanoclay/polymer composite. Compos Part B Eng 42:1708–1712
135. Hassanzadeh-Aghdam MK, Ansari R, Mahmoodi MJ (2019) Micromechanical analysis of the elastic response of glass-epoxy hybrid composites containing silica nanoparticles. Mech Adv Mater Struct 26:1920–1934
136. Dinesh T, Kadirvel A, Hariharan P (2019) Role of nano-silica in tensile fatigue, fracture toughness and low-velocity impact behaviour of acid-treated pineapple fibre/stainless steel wire mesh-reinforced epoxy hybrid composite. Mater Res Express 6:109–127
137. Huang CY, Tsai JL (2015) Characterizing vibration damping response of composite laminates containing silica nanoparticles and rubber particles. J Compos Mater 49:545–557
138. Gokuldass R, Ramesh R (2019) Mechanical strength behavior of hybrid composites tailored by glass/kevlar fibre-reinforced in nano-silica and micro-rubber blended epoxy. Silicon 11:2731–2739
139. Prabu VA, Kumaran ST, Uthayakumar M, Manikandan V (2018) Influence of redmud particle hybridization in banana/sisal and sisal/glass composites. Part Sci Technol 36:402–407
140. Pedrazzoli KD, Pegoretti A (2014) Hybridization of short glass fiber polypropylene composites with nanosilica and graphite nanoplatelets. J Reinf Plast Compos 33:1682–1695
141. Reza Khalili SM, Farsani RE, Soleimani N, Hedayatnasab Z (2014) Charpy impact behavior of clay/basalt fiber-reinforced polypropylene nanocomposites at various temperatures. J Thermoplast Compos Mater 29:1416–1428
142. Ahmadi M, Ansari R, Rouhi S (2019) Fracture behavior of the carbon nanotube/carbon fiber/polymer multiscale composites under bending test–a stochastic finite element method. Mech Adv Mater Struct 26:1169–1117
143. Georgantzinos SK, Giannopoulos GI, Anifantis NK (2009) Investigation of stress-strain behavior of single walled carbon nanotube/rubber composites by a multi-scale finite element method. Theor Appl Fract Mech 52:158–164
144. Malekimoghadam R, Icardi U (2019) Prediction of mechanical properties of carbon nanotube-carbon fiber reinforced hybrid composites using multi-scale finite element modelling. Compos Part B Eng 177:1–15
145. Adamu M, Rahman MR, Hamdan S (2019) Formulation optimization and characterization of bamboo/polyvinyl alcohol/clay nanocomposite by response surface methodology. Compos Part B Eng 176:1–9
146. Balakrishnan S, Krishnaraj C, Raajeshkrishna CR (2019) Mechanical characterization of pineapple, watermelon peel nanoparticles reinforced carbon, jute fabric and its hybrid epoxy composites. Mater Res Express 6:2053–2062
147. Shahroze RM, Ishak MR, Salit MS, Leman Z, Chandrasekar M, Munawar NS, Asim M (2019) Sugar palm fiber/polyester nanocomposites: influence of adding nanoclay fillers on thermal, dynamic mechanical and physical properties. J Vinyl Addit Technol 26:236–243

148. Jawahar P, Balasubramanian M (2006) Influence of nanosize clay platelets on the mechanical properties of glass fiber reinforced polyester composites. J Nanosci Nanotechnol 6:3973–3981
149. Zhu HG, Liu MY, Yuen RK, Leung CK, Kim JK (2014) Thermal performance and flame retardancy studies of vinyl ester and glass fiber reinforced plastic composites containing nanoclay. J Compos Mater 48:165–177
150. Shaker K, Nawab Y, Saouab A, Ashraf M, Khan AN (2018) Effect of silica particle loading on shape distortion in glass/vinyl ester-laminated composite plates. J Text Inst 109:656–664
151. Chandradass J, Ramesh Kumar M, Velmurugan R (2008) Effect of clay dispersion on mechanical, thermal and vibration properties of glass fiber-reinforced vinyl ester composites. J Reinf Plast Compos 27:1585–1601
152. Raghavendra N, Murthy HN, Mahesh KV, Mylarappa M, Ashik KP, Siddeswara DMK, Krishna M (2017) Effect of nanoclays on the performance of mechanical, thermal and flammability of Vinylester based nanocomposites. Mater Today Proc 4:12109–12117
153. Hussain F, Dean D, Haque A, Shamsuzzoha AM (2005) S2-Glass/vinylester polymer nanocomposites: manufacturing, structures, thermal and mechanical properties. J Adv Mater 37:16–27
154. Park JY, Zureick AH (2005) Effect of filler and void content on mechanical properties of pultruded composite materials under shear loading. Polym Compos 26:181–192
155. Kim SH, Park SJ (2020) Effect of graphene oxide on interfacial interactions and fracture toughness of basalt fiber-reinforced epoxy composites. J Nanosci Nanotechnol 20:6760–6767

Chapter 3
Properties of Natural Fiber-Reinforced Epoxy Hybrid Composites

Abstract Natural fibers have many advantages over synthetic fibers due to their lightness, low cost, biodegradability, and abundance in nature. The demand for natural fiber hybrid composites in various applications has increased recently, because of its promising mechanical properties. In this chapter, the mechanical and wettability properties of reinforced natural fiber epoxy resin hybrid composites were investigated (Atmakuri et al. in Polymers 12(12):1–15, (2020) [1]). The main aim of this chapter is the fabrication of hybrid composites and exploit its importance over individual fiber composites. The composites were fabricated based on the rule of hybridization mixture (0.4 wf) of two fibers using sets of either hemp and flax or banana and pineapple, each set with 40 wt%, as well as four single fiber composites, 40 wt% each, as reinforcement and epoxy resin as matrix material. A total of two sets (hemp/flax and banana/pineapple) of hybrid composites were fabricated by using a hand layup technique. One set as 40H/0F, 25H/15F, 20H/20F, 15H/25F, 0H/40F, and the second one as 40B/0P, 25B/15P, 20B/20P, 15B/25P, 0B/40P weight fraction ratios. The fabricated composites were allowed for testing to examine its mechanical, wettability, and moisture properties. It has been observed that, in both cases, hybrid composites showed improved mechanical properties when compared to the individual fiber composites. The wettability test was carried out by using the contact angle measurement technique. All composites in both cases, hybrid or single showed contact angle less than 90°, which is associated with the composite hydrophilic surface properties. The moisture analysis stated that all the composites responded for moisture absorption up to 96 h and then remained constant in both cases. Hybrid composites absorbed less moisture than individual fiber composites.

3.1 Introduction

Recently, a growing number of researchers have expressed interest in natural fiber hybrid composites due to their abundance, lightweight nature, cost-effectiveness, high specific properties, and environmental sustainability. Natural fibers are increasingly being utilized as reinforcement materials in polymer matrix composites, often

A. Atmakuri et al., *Fiber-Based Hybrid Composites for Engineering*, Springer Series on Polymer and Composite Materials, https://doi.org/10.1007/978-3-031-93144-4_3

serving as substitutes for synthetic (manmade) fibers. Hybrid fiber composites are composed of two or more reinforcing agents alongside a uniform matrix material, a process termed hybridization [2]. The matrix functions as a binding substance, contributing additional strength to the composite [3]. Composites are categorized according to both matrix and reinforcement materials. Based on matrix material, composites are divided into metal matrix composites (MMCs), ceramic-based composites (CMCs), and organic matrix composites (OMCs). In terms of reinforcement material, composites are classified as laminar composites, particulate composites, and fiber-reinforced composites (FRCs) [4–6]. Within fiber-reinforced composites, fiber material acts as a reinforcement component, providing stiffness. The matrix material not only serves as a binding agent but also redistributes external forces (loads) exerted on the composite toward the fibers. The majority of the external load is borne by the reinforcement materials, while the matrix material supports a minor portion. Fibers are categorized into natural and synthetic fibers. Examples of natural fibers include abaca, bamboo, banana, cotton, coir, flax, hemp, jute, kenaf, pineapple, and sisal, whereas boron, carbon, ceramic, and glass represent synthetic fibers, also referred to as manmade fibers. Synthetic fiber composites exhibit numerous drawbacks compared to natural fiber composites, such as rapid combustion, lack of environmental friendliness, and discomfort due to being non-skin-friendly, making them unsuitable for prolonged use [7]. In contrast, natural fibers are biodegradable, cost-effective, eco-friendly, lightweight, widely available in nature, and renewable. The biodegradability of natural fibers significantly contributes to maintaining a healthy environment coupled with cost-effectiveness and high specific performance, garnering substantial economic interest from various industries. The physical and mechanical properties of natural fibers and their composites continue to be an area of ongoing investigation.

Hybrid composites are synthesized to preserve the benefits of their constituents while mitigating certain limitations. The concept of hybridization represents an effective strategy in plant fibers to address the deficiency in ductility of the composites [8]. Hybridization is defined as the fabrication of composites incorporating two or more reinforcements, integrated within the same matrix (resin) material at a 0.4 weight fraction ratio. Multiple researchers have reported that the hybridization of different fibers across varying weight fractions results in a gradual increase in mechanical and physical properties, leading to cost-effective composite materials. A primary advantage is that the amalgamation of fibers can be achieved through various methods, including intermixed continuous fibers, intermixed discontinuous fibers, intermingled particulate fibers, aligned short fibers, and sandwich layers [9, 10]. The hybridization of natural fibers has resulted in composites exhibiting significantly enhanced mechanical properties. A recent survey highlighted the growing popularity of fabricating natural fiber-reinforced polymer composites for commercial applications, including construction, automotive, military goods, sports equipment, aerospace, and shipping industries. Fernanda Luz et al. [11] conducted studies on composites using natural fibers and conventional materials in hard armor applications. Their findings indicated that natural fiber composites in armor plates can reduce the thickness of the hard armor plate. The experimental results demonstrated that natural fiber-reinforced composites have the potential to replace ceramic-based composites. Neuba et al. [12] researched the mechanical, thermal, and ballistic properties of novel epoxy composites reinforced with Cyperus malaccensis sedge fiber. Their findings revealed an improvement in mechanical properties compared to neat epoxy composites. Yorseng et al. [13] conducted weathering studies on kenaf and sisal fiber reinforced hybrid composites for semi-structural applications. The findings showed that the mechanical and thermal properties of natural fiber hybrid composites were more enhanced than those of pure epoxy composites, both before and after weathering conditions. These results suggest a significant advantage in utilizing bio-based composites in semi-structural applications.

Chemical composition is a critical factor in the properties of natural fibers, which vary significantly among different fiber types. The performance of such fibers is predominantly dependent on the aspect ratio and cellulose crystallinity of the material. Cellulose is a primary constituent of plant fibers, significantly influencing moisture absorption. In addition to cellulose, hemicellulose and lignin are vital components. Cellulose constitutes up to 55% of the total composition of plant fibers. Lignin functions as a cementing agent, influencing surface morphology and properties [14]. Plant fibers and their derivatives form the basis for hybrid composites, which are prevalent lignocellulosic materials due to their superior mechanical properties. In India, banana and pineapple are extensively cultivated plants providing leaf fibers for composite applications. The use of these leaf fiber composites and their by-products has broadened to encompass various commercial applications. Additionally, bast fiber composites, particularly from hemp and flax, are gaining popularity. These fibers are predominantly cultivated in European nations [15]. Researchers are increasingly interested in these materials due to their mechanical performance and suitability for lightweight applications. According to the European Commission, utilizing waste raw materials derived from the processing of bast fibers like hemp and flax supports the sustainable exploitation of plant resources [16]. Beyond the reinforcement materials, matrix materials also play a crucial role in hybrid composites. The structure and surface morphology characteristics of composites are influenced by the resin material used. Matrix materials are categorized into thermosetting and thermoplastic materials based on processing techniques. They function as load distributors within a composite subjected to external forces [17]. Thermosetting polymers are highly regarded due to their unique ability to establish three-dimensional bonds upon curing, distinguishing them from plastic polymer materials. They are characterized by exceptional mechanical performance, thermal stability, and resistance to creep [18]. Thermoset polymers are more prevalently utilized as resin materials in composite fabrication than thermoplastic polymers. Thermosetting materials exhibit superior mechanical properties, thermal stability, and are facile in terms of processing. The processing of thermosets is straightforward and does not necessitate supplementary heat treatment for composite curing. In contrast to thermoplastic materials, they are not recyclable. Polystyrene, polyethylene, polyvinyl chloride, and natural rubber are examples of thermoplastic resins. Conversely, epoxy, phenolic, polyester, vinyl ester, urea–formaldehyde, and unsaturated polyester serve as examples of thermoset resins. There is a notable interest in epoxy resin among researchers owing to its exceptional mechanical properties, rapid curing time, durability, availability, and economical cost [19]. Epoxy resin, accompanied by a hardener, is employed as a matrix material in the current research.

Numerous researchers have determined that reinforcement with natural fibers within an epoxy resin matrix yields enhanced properties. Sapuan et al. [20] conducted a study on the mechanical properties of banana fiber-reinforced epoxy resin composites. Their findings indicated that these banana fiber composites exhibited improvements in flexural strength, durability, and stiffness. Al Rashid et al. [21] explored the application of banana fiber-reinforced hybrid composites in the sports industry, noting that these composites, due to their extensive mechanical properties, are a viable alternative to glass fiber composites in this sector. Karimzadeh et al. [22] examined the influence of stacking sequence on the mechanical properties and moisture absorption characteristics of hybrid pineapple/glass fiber composites, finding that hybrid composites possess superior flexural and tensile strengths compared to pure composites, and that pure pineapple absorbs a higher percentage of water than hybrid composites. Shivamurthy et al. [23] investigated the sliding wear, moisture absorption, flammability, and mechanical properties of banana short fiber/$Al(OH)_3$ epoxy composites. Their results demonstrated significant improvements in mechanical properties such as tensile strength and hardness with the incorporation of natural fibers, with these properties being further enhanced by increasing the fiber material to a certain level. Neves et al. [24] examined mechanical properties comparatively between biocomposites of epoxy and polyester matrices reinforced by hemp fiber, revealing that hemp fiber composites with epoxy resin exhibit superior properties to those of polyester composites. The flexural and tensile strengths of epoxy and polyester resin composites were reported as 66.7 and 49.1 MPa, and 50.5 and 25.4 MPa, respectively. Thiagamani et al. investigated the mechanical, moisture absorption, and swelling behavior of hemp/sisal fiber-reinforced bio-epoxy hybrid composites, specifically focusing on the effects of stacking sequences [25]. The experimental findings suggested that hybrid composites exhibited minimal variation in tensile strength with alterations in the stacking pattern. Notably, hybrid composites demonstrated mechanical properties that were 40% superior to those of individual fiber composites. Moudood et al. [26] conducted an investigation into the mechanical properties of flax fiber reinforced composites under various relative humidities, utilizing experimental, geometric, and displacement potential function methodologies. The experimental data revealed enhancements in both tensile strength and tensile strain in natural hybrid fiber composites. Mohammed Sh. Al-Otaibi et al. [27] focused on the characterization of date palm fiber-reinforced polypropylene matrices. In this research, the authors examined the mechanical, morphological, and thermal properties of date palm fiber composites. The study identified a decrease in tensile properties paired with an improvement in modulus, with these composites exhibiting advanced thermal and morphological attributes compared to neat resin composites. Junio et al. [28] investigated Copernicia prunifera leaf fiber epoxy composites, where results indicated a 40% enhancement in tensile properties and a 69% increase in the elastic modulus when compared to neat epoxy resin composites.

This chapter presents the fabrication of two sets of hybrid composites, each with five distinct weight fractions. These composites are based on epoxy resin and reinforced with four types of natural fibers: hemp (H), flax (F), banana (B), and pineapple (P). The hand lay-up technique was employed for their fabrication. The resulting composites underwent material properties testing, specifically assessing flexural strength, flexural modulus, interlaminar shear strength, contact angle measurement, and moisture absorption. The mechanical performance of the two sets of fabricated laminates was analyzed and their results were compared.

3.2 Materials and Methods

Materials. The reinforcement materials utilized for the preparation of samples comprise banana, flax, hemp, and pineapple natural fibers, along with epoxy resin. The mechanical properties and chemical compositions of these fibers are detailed in Table 3.1. Banana and pineapple fibers are categorized as leaf fibers, which are abundantly available in nature, particularly in India. In contrast, hemp and flax fibers are identified as bast fibers, obtained from the phloem of the plants. The epoxy resin, in conjunction with a hardener, served as the matrix material and was procured from Composites 24, Riga, Latvia. Teak wood was employed as the mold material.

Preparation of fibers. The selection and extraction of fibers constitutes a significant portion of this research. Initially, all acquired fibers were subjected to washing with water followed by drying to remove residual moisture. The fibers were thereafter meticulously segregated manually. Subsequently, the fiber laminates underwent a hand retting process to dissociate individual fiber strands through a mechanical combing technique. Post separation, the fibers were examined multiple times using a cotton checking framework to further isolate the filaments. The fibers were then allowed to air dry at ambient room temperature to ensure complete removal of moisture. Finally, the fibers were measured for their weight and length to meet the specified dimensions. The fibers utilized for fabrication purposes are depicted in Fig. 3.1.

Table 3.1 The chemical and mechanical properties of natural fibers [29, 30]

Properties	Banana	Pineapple	Hemp	Flax
Density (g/cm^3)	1.35	1.44	1.48	1.4
Young's modulus (GPa)	27–32	60.82	70	60–80
Elongation (%)	2.6–5.9	2.2	1.6–4.0	1.2–1.6
Cellulose (%)	63–83	12–60	70–74	64–42
Hemi cellulose (%)	6–19	19.5	21–24	16
Lignin (%)	5–10	4.6–11	3.7–5.7	2–2.2
Moisture (%)	10.71	8–11.8	6.2–12	8–12
Microfibrillar angle (°)	11–12	12–14	2–6	5–10

Fig. 3.1 Fibers used for the fabrication

Weight fraction of reinforcements and matrix materials. The weight fraction of the constituents used for the fabrication process was considered based on the hybridization process. In the present work, total two sets of banana(B)/pineapple(P) and hemp(H)/flax(F) composites were fabricated by varying the weight fraction of fibers, such as 40B/0P (40% banana fibers), 25B/15P (25% banana and 15% pineapple), 20B/20P (20% banana and 20% pineapple), 15B/25P (15% banana and 25% pineapple), 0B/40P (40% pineapple fibers), 40H/0F (40% hemp fibers), 25H/15F (25% hemp and 15% flax), 20H/20F (20% hemp and 20% flax), 15H/25F (15% hemp and 25% flax), and 0H/40F (40% flax fibers).

Preparation of matrix material. Epoxy resin, together with a hardener, was employed as the matrix material. In accordance with the outlined instructions, the appropriate resin solution was achieved by maintaining a weight ratio of resin to hardener at 10:2. The epoxy resin and hardener were placed in a plastic container and agitated using a plastic stirrer for a duration of 2–3 min at ambient temperature, until the mixture attained a uniform coloration. Subsequent to adequate amalgamation of the hardener with the epoxy, the resin solution was administered onto the fibers. The mechanical characteristics of the epoxy resin and hardener are presented in Table 3.2.

Preparation of Mold. The mold utilized in this study was constructed from teak wood, as depicted in Fig. 3.2a. The mold has dimensions of 150 mm × 150 mm × 12 mm. Typically, the matrix material adheres to the surface of the mold. To prevent

Table 3.2 Mechanical properties of epoxy resin and hardener [31]

Property	Epoxy Resin	Hardener
Density (g/cm^3)	1.15–1.18	0.97–0.99
Specific gravity	1.14 ± 0.1	1.02 ± 0.1
Color	Colorless	Clear or brown
Viscosity at 25 °C (MPa)	550 ± 50	600
Pot life (min) at 23 °C	30 ± 10	30 ± 10
Mixing proportion	10	2
Gel time (h) at 23 °C	24–36	24–36

(a)

(b)

Fig. 3.2 Mold material for fabrication purpose (**a**); roller on the composite surface (**b**)

this adhesion, baking paper was used to cover the mold, as it does not adversely affect mechanical properties. Upon placement of the fibers within the mold, resin material was applied, and a roller was employed, as illustrated in Fig. 3.2b, to ensure uniform distribution.

Composite fabrication. Hybrid composites of banana/pineapple and hemp/flax fibers were fabricated utilizing hand lay-up techniques. Initially, the mold materials were meticulously cleaned prior to the placement of fibers within them. The mold was then covered with baking paper, and the fibers were arranged in a bed-like formation before the application of resin. Following the pouring of epoxy resin into the mold, it was compressed using a roller for a few minutes to ensure even distribution of the epoxy throughout the mold corners, thereby mitigating the occurrence of gaps and holes in the composite. A uniform load of 6KN was applied to the surface of the composite, which was then allowed to cure for a duration of 24–28 h. For each specified weight fraction, five samples were prepared to enable the calculation of average values and to ensure the specimens were cut in accordance with ASTM standards. The fabricated samples of hemp/flax and banana/pineapple hybrid composites are shown in Fig. 3.3a, b. The weight fractions of the composites are 40B/0P, 25B/15P, 20B/20P, 15B/25P, 0B/40P, 40H/0F, 25H/15F, 20H/20F, 15H/25F, and 0H/40F.

3.2.1 *Testing of Composites*

Flexural strength test. The flexural tests were performed to find the mechanical properties of the composite materials. This test is used to calculate the maximum stress and strain for the addition of external load. The specimens were tested on a Tinius Olsen H10K machine, Horsham, United States of America at a constant strain rate of 0.10 mm/min and speed as 20 mm/min. The force accuracy as 0.5% of applied load and 0.001 mm/min speed resolution. All the samples were tested at room temperature, and every time, five specimens were tested for each composite to

Fig. 3.3 Hemp and flax fiber hybrid composites (a); banana and pineapple fiber hybrid composites (b)

get the average values. The specimens were tested as per ASTM D 790 standards. The dimensions of the samples are taken as 100 mm total length, 80 mm span length, 20 mm, and 5 mm thickness. The samples were allowed for three-point bending tests. The apparatus used for the flexural test is shown in Fig. 3.4a. The strength of the composites was calculated by using the following relation.

$$\sigma_f = \frac{3PL}{2bd^2} \text{ MPa} \tag{3.1}$$

where σ_f is flexural strength, P is maximum load, L is length of the composites, b is width, and d is thickness.

Flexural Modulus Test. The flexural modulus of composite material is defined as the ability for that composite to deform. It is calculated from the slope of the stress and displacement curve. It is also called a tangent modulus and modulus of elasticity. The following relation was used to find out the flexural modulus.

$$E_B = \frac{L^3 m}{4bd^3} \text{ MPa} \tag{3.2}$$

where E_B is flexural modulus, L is length, m is slope of stress–strain curve, b is width, and d is thickness.

Interlaminar Shear Strength. Shear tests of various types are widely used in the polymers industry to find the strength of the reinforcement to matrix materials. The interlaminar shear strength is the failure shear occurred in a composite material when the transverse load applied to it. The samples were taken for a notch cut near to the center on both sides of the specimen before going for the testing. All the specimens were allowed for testing on the Tinius Olsen H10K apparatus. The specimens were tested as per ASTM D-3846-02 standards. The dimensions of the samples are taken as 100 mm total length, 79.6 mm span length, 12.7 mm width, 5 mm thickness, 2.5 mm notch depth, and 6.4 mm distance between the notches. The following relation was used to calculate the shear strength of the composites. The apparatus used for this

(a) (b)

Fig. 3.4 Three-point bending test for flexural properties on Tinius Olsen (**a**); interlaminar shear strength test on Tinus Olsen (**b**)

test is shown in Fig. 3.4b.

$$\sigma_S = \frac{P}{A}\ \text{MPa} \tag{3.3}$$

where P is maximum load and A is area.

Moisture Absorption Test. This test is used to find out the rate of water (moisture) absorbed by the composite when it is exposed to the wet medium. To evaluate the absorption rate, the composite sample is placed in a container full of distilled water. These samples were placed in an oven for 20 min at 60 °C for the heat treatment process, before placing it in the container to eliminate the moisture in it. The specimens were tested as ASTM D-570. The dimensions of the samples are taken as 76 mm length, 5 mm thickness, and 20 mm width. The % of weight gain by the composites were calculated in successive time intervals for five days. The following relation was used to calculate the moisture absorption of composite samples.

$$\%\ Weightgain = \frac{Finalweight - Initialweight}{Initialweight} \times 100 \tag{3.4}$$

Contact Angle Measurement Test. The contact angle measurement technique is used to find out the wettability of the composite surfaces. The wetting property

Fig. 3.5 Contact angle measurement apparatus

defines the ability of the liquid constituent to hold the interaction when it contacts the solid surface. It was found that the contact angle is an attractive method to observe the behavior of the liquid on a composite surface. It is used to find the hydrophobicity of the composite surfaces. The specimens were tested as ASTM D-7334. The dimensions of the specimens are taken as 80 mm length, 20 mm width, and 5 mm thickness. A total of five samples were tested for each composite. In this method, a water droplet is placed on the composite surface with the help of a glass pipet. The volume of the water droplet kept constants for all the composites. The experimental setup used for contact angle measurement is shown in Fig. 3.5, and it consists of an image processor, a moveable holder, camera, optical lenses, test sample holder, and a composite placed on a holder rest.

3.3 Results and Discussion

Flexural properties. The analytical results for banana-pineapple and hemp-flax fabricated hybrid composites, obtained from flexural properties testing, are mentioned in the following Table 3.3.

Figure 3.6a, b show the relation between break load and weight fraction of various composites. In the first set of composites pure hemp (40H/0F) and pure flax (0H/40F) composites showed the least break load and hybrid composites showed improved break load. Additionally, 25H/15F showed the highest break load as 364.9 N at the breakpoint of the composite. In the second set of composites pure banana (40B/0P) and 20B/20P composites exhibited low break load as 135 N and 202.6 N, whereas 25B/15P showed the highest break load as 258.7 N. Hemp and flax fiber hybrid composites showed superior properties than banana and pineapple composites.

The flexural properties of various weight fractions of composites are given in Fig. 3.7a, b. Among all the composites, the 25H/15F hybrid composite has the highest flexural strength at 87.57 MPa, and pure banana (40B/0P) has the least strength at 39.36 MPa. Hemp and flax fiber composites exhibited a range between

Table 3.3 Flexural strength properties

Composites	Break load (N)	Flexural strength (MPa)	Flexural modulus (GPa)
40H/0F	272.5	65.39	2.68
0H/40F	262.9	63.10	2.75
15H/25F	331.8	79.64	2.97
20H/20F	288.9	69.34	2.92
25H/15F	364.9	87.57	3.43
40B/0P	135.0	39.36	0.96
0B/40P	214.3	46.70	1.62
15B/25P	206.1	53.76	1.19
20B/20P	202.6	45.84	1.63
25B/15P	258.7	68.54	2.02

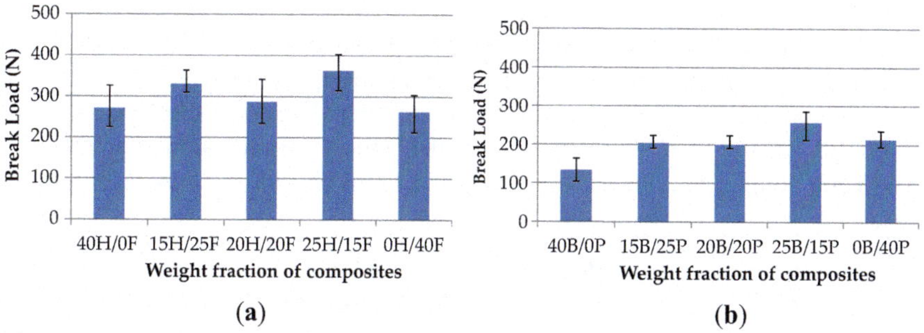

Fig. 3.6 Break load versus weight fraction of hemp/flax (H/)F composites (**a**) and banana/pineapple (B/P) composites (**b**)

63 and 88 MPa, whereas banana and pineapple composites showed in between 38 and 69 MPa. This is due to: (1) the hybridization impact for the fibers contributing superior flexural strength to the composites in both the cases; (2) the poor bonding between fiber reinforcement and matrix material in the pure composites leading to agglomeration, hence, the loss of strength, and fabrication errors might be the reason for lowering the strength in pure composites. The obtained results are proved that the hybridization of natural fibers gives superior results when compared to the individual fiber composites and natural fibers combined with synthetic fibers. It was proved from the previous results, giridharan worked on preparation and property evaluation of glass/ramie fibers reinforced epoxy hybrid composites [32]. The results indicated that the glass and ramie hybrid composites have flexural strength properties range of 50 and 65 MPa, which is less than hemp and flax hybrid composites. Singh et al. [33] investigated the tensile and flexural properties of hemp fiber reinforced virgin-recycled HDPE matrix composites. Hemp fiber composites were fabricated

by varying the weight fraction of fiber and matrix material. The results showed that hemp fiber composites have flexural properties ranging from 16 to 23 MPa, which is far less than the hemp/flax and banana/pineapple hybrid composites. Bazan et al. [34] investigated the mechanical, thermal, and ageing properties of bio-based polyethylene natural fiber composites. The results indicated that flexural properties were increased by increasing the fiber content and a similar trend followed up to 40% of fiber fraction and then decreased. It was noted that the properties varied between 20 and 45 MPa. Sinha et al. [35] investigated the mechanical properties of natural fiber polymer composites. The results stated that coconut fiber epoxy composites showed flexural strength as 64.6 MPa. Raghuram et al. [36] worked on characteristics of treated natural fiber epoxy composites with time-variant. The result stated that natural fiber epoxy-based composites showed flexural strength values between 33.35 and 70 MPa, which is better than the glass fiber blend results. Kumar [37] worked on a dataset on mechanical properties of natural fiber reinforced polyester composites for engineering applications. It was observed from the results that sisal, coconut coir fiber composites exhibited flexural strength ranging from 26.2 to 40.3 MPa. By considering the previous results.

Figure 3.8a, b show the trend of the flexural modulus of various composites. It was observed from the results that banana and pineapple composites showed lower flexural modulus than hemp and flax composites. Pure banana (40B/0P) has the least among all, and the 15H/25F composite has the highest flexural modulus at 2.97 GPa. All hybrid composites showed the flexural modulus close to each other in both the sets.

Interlaminar Shear Strength. The data presented in Table 3.4 show that the interlaminar shear strength properties for all the composites. The tests were performed to observe the failure shear of the composites. The results were considered when the failure shear occurred in between the notches.

Figure 3.9a, b show the interlaminar shear strength with respect to the weight fraction of all the composites. In both cases, pure composites showed the highest shear strength values and hybrid composites showed intermediate results. The pure hemp (40H/0F) composite showed the highest shear strength at 13.01 MPa, and the

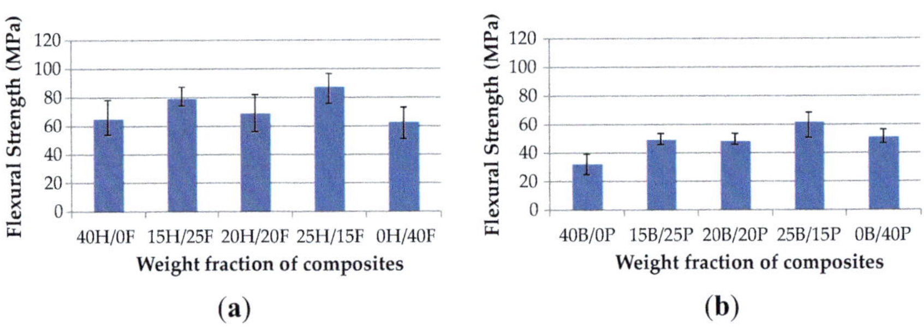

Fig. 3.7 Flexural strength versus weight fraction of H/F composites (**a**) and B/P composites (**b**)

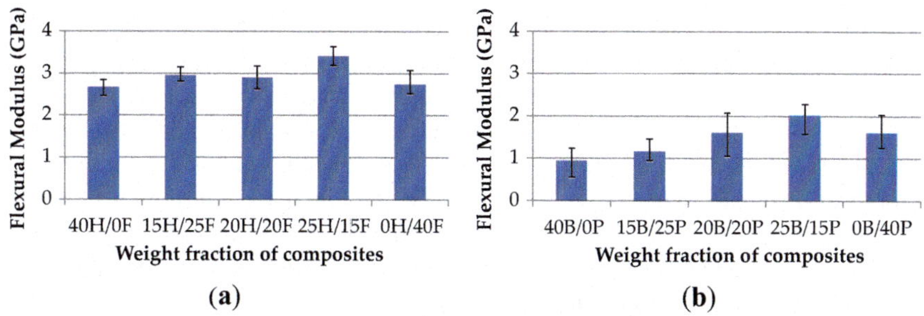

Fig. 3.8 Flexural modulus versus weight fraction of H/F composites (**a**) B/P composites (**b**)

Table 3.4 Interlaminar shear strength properties

Composites	Break load (N)	Interlaminar shear strength (MPa)
40H/0F	513.29	6.32
0H/40F	1057.67	13.01
15H/25F	568.15	6.99
20H/20F	682.27	8.39
25H/15F	540.67	6.65
40B/0P	246.30	3.03
0B/40P	594.77	7.32
15B/25P	403.53	4.96
20B/20P	462.42	5.69
25B/15P	518.70	6.38

pure banana (40B/0P) composite showed the least shear strength at 3.03 MPa. The hybridization effect is negligible, because there is no such variation in results in both.

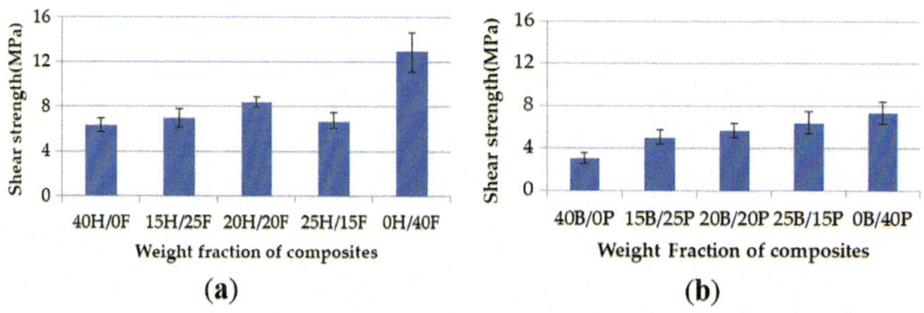

Fig. 3.9 Shear strength versus weight fraction of H/F composites (**a**) and B/P composites (**b**)

Moisture Analysis. The amount of water absorbed by composite material mainly depends on the fiber content, immersion temperature, area of contact to water, and void content. The results were calculated at regular intervals of time for five days.

Figures 3.10 and 3.11 show the percentage of weight gain in all fabricated composites. It was observed from the results that the percentage of weight gain in the composite increases with the increase of time. The increase rate remains constant after 96 h of time in both cases, which means that the composite reaches an equilibrium state after 96 h. Pure hemp and pure banana composites showed the highest and pure flax and pure pineapple the lowest values of absorption rate, whereas hybrid composites showed intermediate results. This can be attributed to: (1) the presence of hydroxyl groups in the cellulose structure, which attracts more water and binds the cells through hydrogen bonding; (2) the void content (porosity), which also attracts water uptake in the composite due to the fabrication errors.

Contact Angle Measurement. The wettability properties or hydrophilic nature of the composites are calculated by using the contact angle measure analysis. The contact angle was measured with the image processing apparatus as shown in Fig. 3.5. All the values are measured (recorded) in a dark place and for each composite five samples were tested. The contact angle of the composites was measured and with deviation in each value presented in Fig. 3.12a, b.

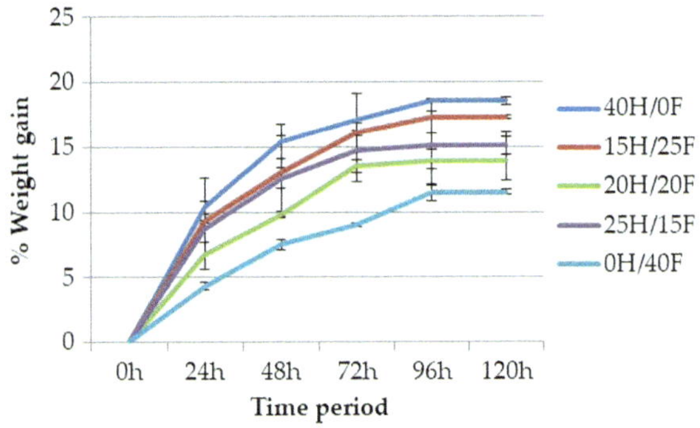

Fig. 3.10 Percentage weight gain versus time period for hemp and flax composites

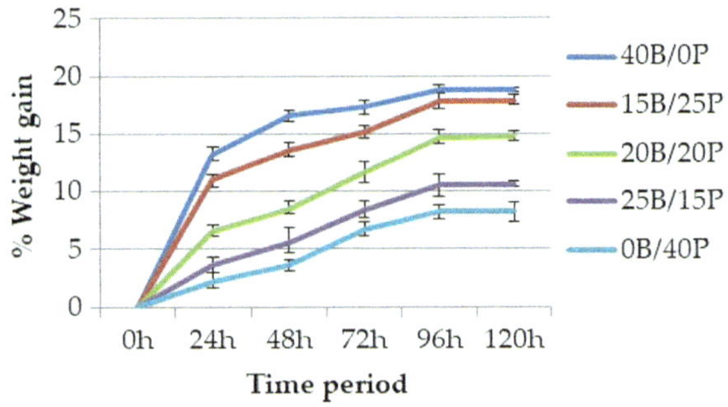

Fig. 3.11 Percentage weight gain versus time period for banana and pineapple composites

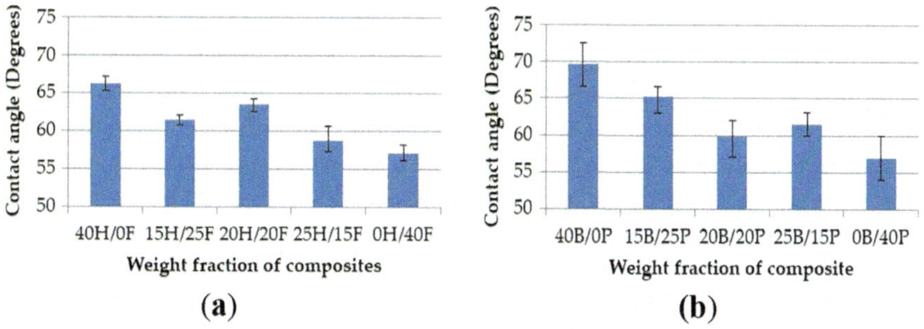

Fig. 3.12 Contact angle versus weight fraction of H/F composites (**a**) and B/P composites (**b**)

In general, the highest contact angle is recorded on rough composite surfaces, and the lowest is recorded on smooth composite surfaces. If the contact angle is less than 90°, then the composite has a hydrophilic nature. If it is more than 90°, then the composite has hydrophobic nature. It was found that all the fabricated composites showed a contact angle of lower than 90°, which means that composites have hydrophilic nature. Hemp and flax fiber hybrid composites exhibited a contact angle between 58° and 70°. Banana and pineapple composites showed between 55° and 75°. Hybrid composites showed an intermediate contact angle in both cases, and pure composites showed the highest contact angle and the lowest contact angle. The is due to both banana and pineapple fibers being cellulose rich fibers, and hybrid composites having a smooth surface finish. From the overall results, it was clear that the hemp and flax fiber composites have a smoother surface finish than the banana and pineapple composites.

3.4 Conclusions

Natural fiber reinforced epoxy resin hybrid fiber composites with varying weight fractions were fabricated and characterized. Two sets of composites (H/F and B/P) were constructed according to the rule of hybridization mixtures, utilizing the hand lay-up technique for comparative analysis. In both instances, hybrid composites exhibited enhanced mechanical properties relative to pure composites. Mechanical properties testing of the first set of composites (hemp and flax) revealed that the hybrid composites possessed superior flexural properties when compared to both pure hemp and pure flax composites. Specifically, the 25H/15F composite demonstrated a flexural strength and flexural modulus of 87.57 MPa and 3.43 GPa, respectively. In contrast, the pure flax composite exhibited the lowest flexural strength and modulus at 63.10 MPa and 2.75 GPa, respectively. In the second set of composites (banana and pineapple), it was determined that the 25B/15P composite exhibited the highest flexural strength and modulus at 68.54 MPa and 2.02 GPa, respectively, whereas the

40B/0P composite displayed the lowest values at 39.36 MPa and 0.96 GPa, respectively. In both scenarios, hybrid composites demonstrated intermediate interlaminar shear strength, with pure composites showing the extremes of highest and lowest values. Contact angle measurement analysis indicated that all the composites exhibited contact angles of less than 90°, signifying hydrophilic surface properties. There was no notable variation in contact angle results among all composites; however, pure composites presented the highest contact angles in both cases, indicating a rougher surface finish. Moisture analysis results revealed that all composites absorbed water when submerged in a water container, with percentage weight gain stabilizing after 96 h, thereby indicating that the composites achieved an equilibrium state. In both cases, hybrid composites absorbed less water compared to pure composites. Taking into account the overall findings, hybrid composites demonstrated enhanced properties compared to pure composites, with hemp and flax fiber composites exhibiting superior properties relative to banana and pineapple composites. Hemp and flax fibers hold potential as reinforcements in composites for structural applications.

References

1. Atmakuri A, Palevicius A, Siddabathula M, Vilkauskas A, Janusas G (2020) Analysis of mechanical and wettability properties of natural fiber-reinforced epoxy hybrid composites. Polymers 12(12):1–15. Basel: MDPI. ISSN 2073-4360. Art no. 2827. https://doi.org/10.3390/polym12122827
2. Del Masto A, Trivaudey F, Guicheret-Retel V, Placet V, Boubakar L (2019) Investigation of the possible origins of the differences in mechanical properties of hemp and flax fibres: a numerical study based on sensitivity analysis. Compos Part A Appl Sci Manuf 124:105488. https://doi.org/10.1016/j.compositesa.2019.105488
3. Charkviani RV, Pavlov AA, Pavlova SA (2017) Interlaminar strength and stiffness of layered composite materials. Procedia Eng. 185:168–172. https://doi.org/10.1016/j.proeng.2017.03.335
4. Rajak DK, Pagar DD, Kumar R, Prusncu CI (2019) Recent progress of reinforcement materials: a comprehensive overview of composite materials. J Mater Res Technol 8:6354–6374. https://doi.org/10.1016/j.jmrt.2019.09.068
5. Sharma AK, Bhandari R, Aherwar A, Rimašauskienė R (2020) Matrix materials used in composites: a comprehensive study. Mater Today Proc 21:1559–1562
6. Kumar A, Singh R, Chaudhary R (2020) Recent progress in production of metal matrix composites by stir casting process: an overview. Mater Today Proc 21:1–5
7. Khalil HPSA, Hanida S, Kang CW, Fuaad NAN (2007) Agro-hybrid composite: the effects on mechanical and physical properties of oil palm fiber (EFB)/glass hybrid reinforced polyester composites. J Reinf Plast Compos 26:203–218. https://doi.org/10.1177/0731684407070027
8. Bhoopathi R, Deepa CK, Sasikala G, Ramesh M (2015) Experimental investigation on mechanical properties of Hemp-Banana-Glass fiber reinforced composites. Appl Mech Mater 766–767:167–172. https://doi.org/10.4028/www.scientific.net/amm.766-767.167
9. Swolfs Y, Crauwels L, Van Breda E, Gorbatikh L, Hine P, Ward I, Verpoest I (2014) Tensile behaviour of intralayer hybrid composites of carbon fibre and self-reinforced polypropylene. Compos Part A Appl Sci Manuf 59:78–84. https://doi.org/10.1016/j.compositesa.2014.01.001
10. Diao H, Bismarck A, Robinson P, Wisnom MR (2014) Production of continuous intermingled CF/GF hybrid composite via fibre tow spreading technology. In: Proceedings of the 16th

European conference on composite materials (ECCM 16), Seville, Spain, 22–26 June 2014, pp 1–8
11. Da Luz FS, Filho FDCG, Oliveira MS, Nascimento LFC, Monteiro SN (1920) Composites with natural fibers and conventional materials applied in a hard armor: a comparison. Polymers 2020:12. https://doi.org/10.3390/polym12091920
12. Neuba LDM, Pereira Junio RF, Riberiro MP, Souza AT, Lima EDS, Filho FDCG, Figueiredo AB-HdS, Braga FdO, de Azevedo ARG, Monteiro SN (2020) Promising mechanical, thermal, and ballistic properties of novel epoxy composites reinforced with cyperus malaccensis sedge fiber. Polymers 12:1776
13. Yorseng K, Rangappa SM, Pulikkalparambil H, Siengchin S, Parameswaranpillai J (2020) Accelerated weathering studies of kenaf/sisal fiber fabric reinforced fully biobased hybrid bio epoxy composites for semi-structural applications: morphology, thermo-mechanical, water absorption behaviour and surface hydrophobicity. Constr Build Mater 235:117464
14. Wang X, Xie J, Zhang H, Zhang W, An S, Chen S, Luo C (2019) Determining the lignin distribution in plant fiber cell walls based on chemical and biological methods. Cellulose 26:4241–4252. https://doi.org/10.1007/s10570-019-02384-2
15. Bourmaud A, Beaugrand J, Shah DU, Placet V, Baley C (2018) Towards the design of high-performance plant fibre composites. Prog Mater Sci 97:347–408. https://doi.org/10.1016/j.pmatsci.2018.05.005
16. Burczyk H, Grabowska L, Kolodziej J et al (2008) Industrial hemp as a raw material for energy production. J Ind Hemp 13:37–48. https://doi.org/10.1080/15377880801898717
17. Rangappa SM, Madhu P, Jawaid M, Senthamaraikannan P, Senthil S, Pradeep S (2018) Characterization and properties of natural fiber polymer composites: a comprehensive review. J Clean Prod 172:566–581. https://doi.org/10.1016/j.jclepro.2017.10.101
18. Bakare FO, Skrifvars M, Åkesson D, Wang Y, Afshar SJ, Esmaeili N (2014) Synthesis and characterization of bio-based thermosetting resins from lactic acid and glycerol. J Appl Polym Sci 131:1–9. https://doi.org/10.1002/app.40488
19. Mohammed AM, Zainab Hamid AA (2017) Synthesis, characterization, and properties of new unsaturated polyesters resins reinforced with some fillers (Carbon Nano, Nano TiO_2, TiO_2 and ZnO) for composite application. Int J Appl Nat Sci 6:79–100
20. Sapuan S, Leenie A, Harimi M, Beng YK (2006) Mechanical properties of woven banana fibre reinforced epoxy composites. Mater Des 27:689–693. https://doi.org/10.1016/j.matdes.2004.12.016
21. Al Rashid A, Khalid MY, Imran R, Ali U, Koç M (2020) Utilization of Banana fiber-reinforced hybrid composites in the sports industry. Materials 13:3167. https://doi.org/10.3390/ma13143167
22. Karimzadeh A, Yahya MY, Abdullah MN, Wong KJ (2020) Effect of stacking sequence on mechanical properties and moisture absorption characteristic of hybrid PALF/glass fiber composites. Fibers Polym. 21:1583–1593. https://doi.org/10.1007/s12221-020-9640-2
23. Shivamurthy B, Thimmappa BHS, Monteiro J (2020) Sliding wear, mechanical, flammability, and water intake properties of banana short fiber/Al (OH)3/epoxy composites. J Nat Fibers 17:337–345
24. Neves ACC, Rohen LA, Mantovani DP, Carvalho JPRG, Vieira CMF, Lopes FPD, Simonassi NT, da Luz FS, Monteiro SN (2020) Comparative mechanical properties between bio composites of epoxy and polyester matrices reinforced by hemp fiber. J Mater Res Technol 9:1296–1304
25. Thiagamani SMK, Krishnasamy S, Muthukumar C, Tengsuthiwat J, Siengchin S, Ismaile SO (2019) Investigation into mechanical, absorption and swelling behaviour of hemp/sisal fibre reinforced bio epoxy hybrid composites: effects of stacking sequences. Int J Biol Macromol 140:637–646
26. Moudood A, Rahman A, Nayeem MLH, Ochsner A, Islam MM, Gaston F (2020) Mechanical properties of flax fiber-reinforced composites at different relative humidities: experimental, geometric, and displacement potential function approaches. Polym Compos 19:1–11

27. Al-Otaibi MS, Alothman OY, Alrashed MM, Anis A, Naveen J, Jawaid M (2020) Characterization of date palm fiber-reinforced different polypropylene matrices. Polymers 12:597. https://doi.org/10.3390/polym12030597
28. Junio RFP, Nascimento LFC, Neuba LDM, Souza AT, Moura JVB, Filho FDCG, Monteiro SN (2020) Copernicia Prunifera leaf fiber: a promising new reinforcement for epoxy composites. Polymers 12:1–20
29. Mochane MJ, Mokhena TC, Mokhothu TH, Mtibe A, Sadiku ER, Ray SS, Ibrahim ID, Daramola OO (2019) Recent progress on natural fiber hybrid composites for advanced applications: a review. Express Polym Lett 13:159–198. https://doi.org/10.3144/expresspolymlett.2019.15
30. Singh H, Singh T (2019) Effect of fillers of various sizes on mechanical characterization of natural fiber polymer hybrid composites: a review. Mater Today Proc 2013:5345–5350
31. Shivamurthy B, Murthy K, Joseph PC, Kabir R, Bhat KU, Anandhan S (2015) Mechanical properties and sliding wear behaviour of jatropha seed cake waste/epoxy composites. J Mater Cycles Waste Manag 17:144–156
32. Giridharan R (2019) Preparation and property evaluation of Glass/Ramie fibers reinforced epoxy hybrid composites. Compos Part B Eng 167:342–345. https://doi.org/10.1016/j.compositesb.2018.12.049
33. Singh S, Deepak D, Aggarwal L, Gupta VK (2014) Tensile and flexural behaviour of hemp fiber reinforced virgin-recycled HDPE matrix composites. Proc Mater Sci 6:1696–1702
34. Bazan P, Mierzwiński D, Bogucki R, Kuciel S (2020) Bio-based polyethylene composites with natural fiber: mechanical, thermal, and ageing properties. Materials 13:1–19
35. Agnivesh KS, Narang HK, Somnath B (2017) Mechanical properties of natural fiber polymer composites. Polym Eng 37:879–895
36. Raghuram V, Maheswaran R, Shijo JS, Thanesh A (2015) Characteristics of treated natural fiber epoxy composite. Int J Pure Appl Math 119:2079–2088
37. Kumar SS (2020) Dataset on mechanical properties of natural fiber reinforced polyester composites for engineering applications. Data Brief 28:105054. https://doi.org/10.1016/j.dib.2019.105054

Chapter 4
Mechanical Properties of Caryota and Sisal Natural Fibers Based Hybrid Composites

Abstract In recent years, natural fiber reinforced polymer composites have garnered considerable attention compared to synthetic fiber composites due to their numerous advantages, including low cost, lightweight, non-toxicity, non-abrasiveness, and biodegradability. Many researchers have exhibited an interest in employing epoxy resin for composite fabrication over other thermosetting and thermoplastic polymers due to its dimensional stability and mechanical properties. In this chapter, the mechanical and moisture properties of Caryota and sisal fiber-reinforced epoxy resin hybrid composites were examined (Atmakuri et al. in Polymers 13(6):1–19, 2021 [1]). The primary objective of these studies is to develop hybrid composites and highlight their significance over single fiber composites. The Caryota and sisal fiber reinforced epoxy resin composites were fabricated employing the hand lay-up technique. A total of five distinct samples (40C/0S, 25C/15S, 20C/20S, 15C/25S, 0C/40S) were developed based on the principle of hybridization. The samples were subjected to testing to assess their mechanical and moisture properties, and the morphology was analyzed using scanning electron microscopy. It was observed that hybrid composites demonstrated enhanced mechanical properties relative to single fiber (individual fiber) composites. The moisture studies indicated that all composites responded to water absorption, but single fiber composites absorbed more moisture than hybrid composites.

4.1 Introduction

A matrix material is essential for the fabrication of composites, serving as a load distributor among the reinforcement materials when external pressure is applied. Polymer composites are categorized into thermoset and thermoplastic polymer composites based on the matrix material employed [2]. In contrast to thermoplastics, thermosets are not recyclable, as they do not revert to their original state once the resin transitions from a liquid to a solid state following the curing process. Thermoplastics, however, only require heating to assume a new shape. Many researchers have shown a preference for thermoset polymer resins over thermoplastic resins due

A. Atmakuri et al., *Fiber-Based Hybrid Composites for Engineering*, Springer Series on Polymer and Composite Materials, https://doi.org/10.1007/978-3-031-93144-4_4

to their broad range of applications. Among the various thermoset resins, epoxy resins have attained significant popularity because of their adhesive properties, low shrinkage, reduced curing time, and effective moisture permeability resistance [3–7]. With their strong adhesive properties, epoxy resins are highly suitable for bonding a variety of materials, including fibers, steel, plastics, and wood [8, 9]. In order to establish a robust network-like architecture between the reinforcement fibers and the matrix material, it is customary to incorporate a curing agent (hardener) with the epoxy resin. However, the substantial cross-linking density inherent in this configuration contributes to crack propagation, thereby influencing the fracture toughness of the composites. There is a considerable body of research focused on enhancing the structural stability through avenues such as chemical and physical modifications, the integration of nanoparticles, and the inclusion of inert pigments, including but not limited to glass, iron oxide, and basalt particles [10–14]. Composites are systematically categorized based on the matrix material into three distinct types: organic matrix composites, ceramic-based composites, and metal matrix composites [15, 16].

In recent years, natural fiber hybrid composites have gained prominence over synthetic fiber composites, primarily due to their lightweight nature, abundance, economic viability, and high tensile strength. These composites are predominantly derived from botanical, zoological, and mineral sources. Examples of natural (plant-based) fibers include sisal, palm, abaca, bamboo, hemp, flax, banana, and pineapple. Conversely, glass, silica, carbon, and ceramic represent synthetic fibers. The non-biodegradable nature of synthetic fiber composites contributes to environmental degradation through the emission of harmful gases, although they are economically advantageous. In contrast, natural fibers, being widely available in nature, do not pose environmental risks [17–19]. The investigation into the physical and mechanical properties of natural fibers and their composites remains ongoing.

In a general context, composite materials comprise solid mass-bearing elements known as reinforcement, while the softer material is referred to as the matrix. The reinforcement provides rigidity and stiffness, thereby supporting the primary load. The matrix, or binder, maintains the position and orientation of the reinforcement. Essentially, the composition of the composite materials is well-balanced in terms of their individual physical and chemical properties. They offer a combination of attributes that individual materials lack, making them suitable for various applications [20, 21]. Composite materials present advantages over conventional materials due to their superior properties such as impact resistance, elasticity, flexural strength, fatigue resistance, and stiffness. These properties enable designs to be more adaptable. As a result of these advantages, composite materials are widely utilized in mechanical engineering applications such as machine components, thermal control systems, internal combustion engines, electronic packaging, marine constructions, the construction industry, military equipment, sports equipment, and the aerospace industry [22, 23].

The chemical composition is a crucial factor influencing plant-based natural fibers and differs among various fibers. The performance of fibers is primarily contingent upon the aspect ratio and cellulose content within the fiber materials. In the context of plant-based fibers, the constituents of cellulose, hemicellulose, and lignin are integral to determining overall performance [24].

Reinforced hybrid materials are formulated by integrating at least two distinct types of fibers within a distinctive matrix material. The hybridization of two kinds of filaments, each with varying lengths and widths, presents several advantages over the use of individual strands alone within a single polymer composite framework. Most studies focus on the hybridization of natural fibers with glass fibers to enhance material properties [25]. In this context, the fibers serve as the primary load-bearing elements, while the matrix material functions in a supporting capacity [26]. As documented by numerous researchers, the hybridization of diverse fibers at different weight fractions generally leads to a progressive increase in mechanical and physical properties, thereby yielding cost-effective composite materials. A crucial advantage is that fiber mixing can be executed through various methodologies, including intermixed continuous fibers, intermixed discontinuous fibers, intermingled particulate fibers, aligned short fibers, and sandwich layers [27]. Sallam et al. [28] examined the mode II fracture toughness of hybrid fiber reinforced concrete materials. The researchers explored the fracture toughness of four distinct hybridization patterns, revealing an improvement in fracture toughness within the hybrid composites. However, the superior performance is contingent upon the specific hybridization patterns employed. The authors adequately address the reasons for this behavior in the discussion section of their subsequent paper, clarifying potential reader inquiries [29]. Additionally, Sallam et al. [30] investigated the long-term behavior of normal-weight concrete incorporating hybrid nanoparticles under gamma radiation exposure. The experimental findings indicated that incorporating nanoparticles into the hybrid composites resulted in enhanced physical and mechanical properties.

The mechanical performance of epoxy resin-based hybrid composites incorporating natural fibers such as hemp, flax, banana, pineapple, jute, sisal, bamboo, palm, abaca, and okra has been extensively investigated by numerous researchers. Studies have indicated that natural fiber composites serve as conventional replacements for synthetic fiber composites. Alamri et al. [31] examined the mechanical properties and water absorption behavior of recycled cellulose fiber-reinforced epoxy composites. The findings revealed that the mechanical properties of epoxy-based hybrid composites improved with increased fiber content. The impact of absorption on mechanical properties was investigated through SEM and FTIR analysis. Observations were made regarding the porosity and bonding between the matrix material and reinforcement. Sallam et al. [32] explored the tribological and mechanical properties of epoxy reinforced with hybrid nanoparticles. Results demonstrated that the incorporation of nanoparticles into the epoxy resin enhanced the wear resistance of the epoxy; however, there were drawbacks, such as reductions in tensile strength and modulus of elasticity. Colomer-Romero et al. [33] analyzed the mechanical properties of hemp-fiber biocomposites manufactured with biobased and standard epoxy resins. The authors reported that composites using conventional epoxy resins demonstrated superior results, whereas biobased epoxy resins encountered fabrication challenges due to their high viscosity, leading to a reduction in performance. Saba et al. [34] examined the dynamic mechanical properties of oil palm nanofiller kenaf fiber epoxy hybrid composites. It was determined that the integration of nanofillers into the epoxy hybrid composites enhanced both the storage modulus and loss modulus. The authors concluded that hybridization produces environmentally favorable nanocomposites with enhanced damping properties and dynamic modulus. Jawaid et al. [35] explored the impact of oil palm and jute fiber treatment on the mechanical performance of epoxy hybrid composites. The results demonstrated an enhancement in mechanical properties such as flexural and impact strength in modified fiber-reinforced hybrid composites compared to their unprocessed counterparts, attributable to improved fiber/matrix interfacial bonding as confirmed by scanning electron microscopy. Hanan et al. [36] assessed the mechanical performance of oil palm kenaf fiber-reinforced epoxy-based bilayer hybrid composites. Findings revealed that the hybridization of kenaf/palm fiber-reinforced epoxy composites resulted in increased tensile and flexural properties. The interfacial bonding between the matrix and filler material was also improved, as evidenced through SEM analysis. Nonetheless, despite these improvements, poor bonding was detected in the scanning electron microscope analysis, which might be ameliorated by employing chemical treatment of the fibers. Cordeiro et al. [37] investigated the effect of composition and fiber modification on the mechanical and dynamic properties of epoxy resin-based natural fiber peach palm composites. The authors inferred from the findings that these composites exhibited a commendable mechanical response even with a fiber content of 70 wt%. Hanan et al. [38] conducted a study on the characterization of hybrid composites reinforced with oil palm empty fruit bunch woven kenaf fabric in an epoxy matrix. This research endeavor focused on examining the mechanical, physical, and morphological characteristics of hybrid composites. The results indicated enhancements in tensile and flexural properties, while a decrease in impact strength was attributed to inadequate bonding or fabrication discrepancies. The scanning electron microscopy (SEM) analysis adeptly illustrated the distinct failure modes of the tensile fractured samples. The hybridization of palm fibers with kenaf fibers did not yield optimal results, owing to insufficient inter-bonding capabilities. Ramesh et al. [39] examined the mechanical properties of composites comprising hybrid glass/sisal/jute fibers. The results indicated that glass/sisal fiber epoxy resin hybrid composites exhibited superior properties compared to jute/glass fiber composites. The integration of sisal with glass fiber contributed to the enhancement of mechanical properties. However, challenges remain associated with the inclusion of glass (synthetic) fibers, such as issues regarding biodegradability and moisture absorption. To address these concerns, the current study has explored the development of Caryota and sisal natural fiber hybrid composites.

In this study, Caryota (C) and sisal (S) fiber-reinforced epoxy resin hybrid composites, exhibiting varied weight fractions, were synthesized employing the hand lay-up technique. Subsequent to the retting process, the fibers underwent chemical treatment. The synthesized samples were subjected to mechanical evaluations encompassing tensile properties, flexural properties, hardness, impact strength, and moisture characteristics. A scanning electron microscope was utilized to examine the surface morphology, chemical composition, and porosity content of the composite samples. The mechanical performance of the synthesized samples was assessed, and the results underwent comparative analysis.

4.2 Materials and Methods

Caryota and sisal fibers serve as reinforcement materials, while epoxy resin is employed as the matrix material. Sisal fibers are classified as a type of leaf fiber extracted from the sisal plant, whereas Caryota fibers are obtained from fishtail palm trees. Both fibers are abundantly available in nature. These plants are predominantly cultivated in Asian countries due to their requirement for consistently warm weather conditions. The chemical and mechanical properties of both sisal and Caryota fibers are provided in Table 4.1. The thermoset polymer epoxy resin, coupled with a hardener, was utilized as the matrix material. Wood material was employed in the preparation of molds for the fabrication process.

Preparation of Fibers. The extraction of fibers constitutes a critical component of this research. Caryota fibers were procured from a fishtail palm tree using a scraping machine, which is comprised of three rollers: a feed roller, a serrate roller, and a leaf scratching roller. Similarly, sisal fibers were obtained from the leaves of the agave sisalana plant. Following their extraction, both types of fibers were cleansed with still water and dried under sunlight. Subsequently, the fibers underwent a chemical treatment process to enhance the interfacial adhesion with the matrix material. A 2% hydrochloric acid (HCl) solution was prepared, in which each type of fiber was immersed separately for a duration of 3–4 h. Post-treatment, the fibers were again washed with still water and dried at ambient temperature. The fiber laminates were then meticulously separated using mechanical combs manually. Following separation, the fibers were cut into fine pieces with scissors to fit the mold dimensions. The fibers utilized for fabrication are depicted in Fig. 4.1.

Weight Fraction of Materials. The consideration of the weight fraction of reinforcements and matrix materials in the development of hybrid composites was grounded

Table 4.1 The chemical composition and mechanical properties of natural fibers [40, 41]

Property	Sisal fibers	Caryota fibers
Cellulose (%)	50–78	37–47
Hemicellulose (%)	10–14	25–34
Lignin (%)	8–11	18–23
Density (g/cm^3)	1.45	0.7–1.55
Young's modulus (GPa)	9.4–22	1–9
Elongation at break (%)	3–7	2–4.5
Moisture (%)	11	13–15
Microfibrillar angle (°)	20–25	14–18

Fig. 4.1 Sisal fibers **a** Caryota fibers **b** used for the fabrication of composites

in the concept of hybridization. This concept is characterized by the production of composites containing two or more distinct fibers integrated within a uniform resin matrix, maintaining a weight fraction ratio of 0.4 (0.4 wf.). In this study, five distinct composites, each with varying weight fractions, were fabricated, and the corresponding weight proportions are presented in Table 4.2.

Preparation of Matrix Material. Epoxy resin in conjunction with hardener was employed as the matrix material for fabrication purposes. These components were procured from Sigma Aldrich. The weight proportions adhered to a 10:1 ratio as specified by the instructions. The requisite quantities of epoxy and hardener were combined in a plastic container and stirred with a plastic stirrer for 3 to 4 min to ensure a homogenous mixture, then allowed to rest for 30 s. The chemical and mechanical properties of the epoxy resin and hardener are presented in Table 4.3.

Fabrication of Composites. Caryota and Sisal fibers are measured for proportionate weight ratios as discussed above. The fibers were laid uniformly inside the mold before applying any resin to it. After arranging the fibers uniformly, they are compressed for a few minutes in the mold. The compressed fibers were laid over the coat of epoxy resin, ensuring uniform distribution of fibers. The epoxy resin mixture has been poured over the fibers uniformly and compressed for a curing time

Table 4.2 Fiber weight fractions

Composite type	Caryota fiber (%)	Sisal fiber (%)	Total fiber volume (%)	Total resin volume (%)
40C/0S	40	0	40	60
25C/15S	25	15	40	60
20C/20S	20	20	40	60
15C/25S	15	25	40	60
0C/40S	0	40	40	60

Table 4.3 Encapsulates the chemical and mechanical properties of epoxy and hardener [42]

Property	Epoxy resin	Hardener
Type	Araldite LY 556	Ardur HY 951
Mixing proportion	10	1
Color	Pale	Brown
Specific gravity	1.14	1.02
Density (g/cm^3)	1.15–1.18	0.97–0.99
Viscosity at 25 °C (MPa)	550	600
Curing time (h) at 23 °C	24–32	24–32
Pot life (min) at 23 °C	35	35

of 24 h, with a constant load of 5 kg. All samples are fabricated according to ASTM D 638 M standards. The fabricated samples of Sisal/Caryota fiber hybrid composites are shown in Fig. 4.2.

Tensile Properties. Tensile tests were performed to find out the in-plane tensile properties of polymer composites fabricated with matrix and reinforcements. In a broad sense, the tensile test is a measurement of the ability of a material to withstand external forces that tend to pull apart and to what extent the material stretches before breaking. The composite samples were tested on Tinus Olsen H10KT at a constant speed. A total of five samples were tested for each composite and the dimensions were considered as per ASTM D 3039 standards. The experimental setup used for the tensile test is shown in Fig. 4.3a. The tensile strength of composites was calculated by using the following relation.

$$\sigma_t = \frac{P}{A} \text{MPa} \tag{4.1}$$

where σ_t = tensile strength, P = maximum load, and A = cross sectional area.

The tensile modulus of a composite is the ratio of stress to elastic strain when the material is subjected to tension. The following relation was used to calculate the

Fig. 4.2 Sisal and Caryota fiber hybrid composites

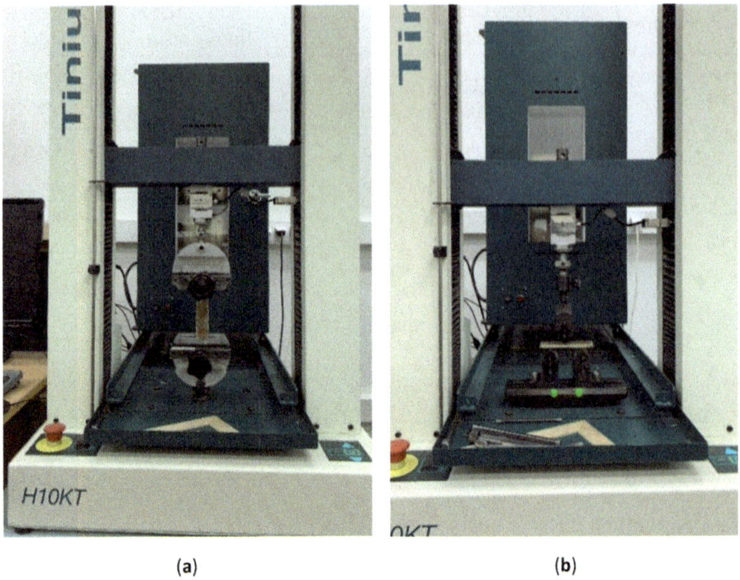

Fig. 4.3 Tensile test **a** flexural test (three-point bending test) **b** on Tinus Olsen H10KT

tensile modulus.

$$E = \frac{FL}{Ae}\text{MPa} \tag{4.2}$$

where E = tensile modulus, F = maximum load, A = cross sectional area, and e = change in dimension.

Flexural Properties. The flexural tests were performed to find out the maximum stress and strain when the composites are subjected to external loading. To find out the flexural strength, composite samples are allowed for 3-point bending tests on Tinus Olsen at a constant strain rate of 0.10 mm/min and a crosshead speed of 20 mm/min. The dimensions of samples were considered as per ASTM D 790 standards. The experimental setup used for the flexural test is shown in Fig. 3b. The flexural strength was calculated by using the following relation.

$$\sigma_f = \frac{3PL}{2bd^2}\text{MPa} \tag{4.3}$$

where σ_f = flexural strength, P = maximum load, L = Span length, b = Width, and d = thickness.

Flexural modulus or tangent modulus is defined as the capability of a composite sample to deform. It is calculated from the tangent of the stress–strain curve. The following equation was used to calculate the flexural modulus.

$$E_B = \frac{L^3 m}{4bd^3} \text{ MPa} \tag{4.4}$$

where E_B = flexural modulus, L = span length, m = slope, b = width, and d = thickness.

Impact Strength Test. This test is used to define the ability of composite material to withstand the sudden shock loads. This test was carried on a Coesfeld Magnus impact testing machine. The drop tower swung from a set height and which on releasing possess fixed kinetic energy. The test samples were placed like a simply supported beam on the resting position of the impact machine. The tests were conducted as per ASTM D 6110 standards. The experimental setup used for this test is shown in Fig. 4a.

Vickers Hardness Test. This test is used to find the ability of composite material to resist indentation, scratch, or surface abrasion. In this work, the hardness test was carried out on the Micro Vickers hardness testing machine of model DH85 and Daksh made. It has a load range of 0.2–5 kgs and a measuring range of 5–3000 HV. The measuring magnification and observer magnification are given as 500X and 100X respectively. The experimental setup used for this test is shown in Fig. 4b.

Moisture Absorption Test. A Moisture absorption test is used to find out the amount of water absorbed by a composite material when it is exposed to moisture. To find out the rate of absorption, composite samples were placed in a beaker full of distilled water. The sample dimensions were considered as per ASTM D 570 standards. The

(a) (b)

Fig. 4.4 Coesfeld Magnus impact testing system (**a**) Micro Vickers hardness tester (**b**)

readings are noted for continuous-time intervals for one week. A total of five samples were tested for each composite material. The rate of absorption was calculated by using the following relation.

$$\% \, Weight \, gain = \frac{W_2 - W_1}{W_1} \times 100 \tag{4.5}$$

where W_2 = Final weight, W_1 = Initial weight.

Scanning Electron Microscopic Analysis. The failure studies, surface morphology, and chemical composition of the composite samples were evaluated by using the scanning electron microscope (S-3400N from Hitachi) with an energy dispersive X-ray spectrometer (Quad 5040 from Bruker, Billercia, MA, USA).

4.3 Results

Tensile Properties. The analytical results for Caryota and the sisal fiber hybrid composites obtained from tensile tests are mentioned in the following Table 4.4. All the composites' specimens were tested as per the standards and for each composite, a total of five samples were tested to note down the average values. The tensile stress was applied in the longitudinal to the fibers direction and speed was maintained as 2 mm/min. The tests were conducted at 21 °C temperature and relative humidity as 28%. The weight fractions of composites are 40C/0S, 25C/15S, 20C/20S, 15C/25S, 0C/40S.

Figure 4.5 shows the tensile strength behavior for Caryota and the sisal fiber hybrid composites with varying weight fractions. From the results, the tensile strength values of 15C/25S hybrid composites are inferior compared to the other composites. It showed the tensile strength as 38.2 MPa whereas 20C/20S showed 35.4 MPa. The single fiber composites showed the least tensile strength as 22.2 and 25.7 MPa. The hybridization effect is noticed in this case. It was observed that the tensile strength values among the hybrid composites increased with an increase in sisal fiber content in it. The reason for the above can be attributed in two ways. An increase in sisal fiber content increases the strength of the composite. However, higher individual fiber content leads to the agglomeration of fibers, hence loss of strength. Poor bonding

Table 4.4 Tensile strength properties

Composites	Tensile strength (MPa)	Tensile modulus (GPa)
40C/0S	22.2	4.20
25C/15S	28.1	5.92
20C/20S	35.4	6.19
15C/25S	38.2	6.44
0C/40S	25.7	5.66

between matrix and reinforcement material at the interface might be another reason for lower the tensile strength value.

The tensile modulus of various weight fractions of composites is given in the following Fig. 4.6. Among all the composites hybrid composites showed superior properties over single fiber composites and the values are close to each other. Single fiber Caryota composites showed the least tensile modulus among all the composites and 15C/25S composites showed the highest tensile modulus.

Flexural Properties. The data presented in Table 4.5 demonstrates that the flexural properties for all the composites. Flexural tests were performed by using three-point bending tests and the dimensions of the test samples are taken from ASTM standards. The flexural test results include displacement and strength. The tests were conducted at 21 °C temperature and relative humidity as 28%. For each composite, a total of five samples were tested and the values are noted.

Figure 4.7 shows the relation between flexural strength and weight fraction of the fabricated composites. As it can be seen from the below figure that the hybrid composites showed superior flexural properties than single fiber composites. 15C/25S hybrid composites showed the highest flexural strength as 89.16 MPa and single fiber Caryota composites showed the least strength as 64.09 MPa. Hybrid composites exhibited good flexural properties which indicate that the materials have brittle

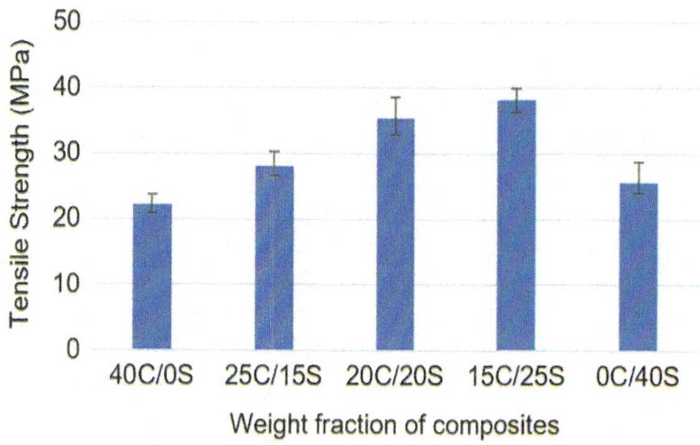

Fig. 4.5 Tensile strength versus weight fraction of C/S composites

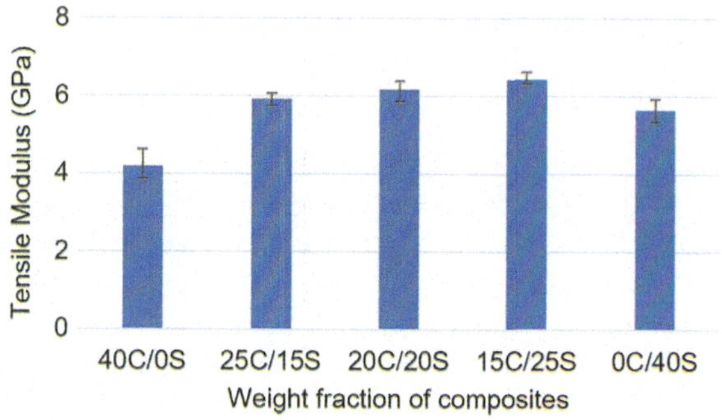

Fig. 4.6 Tensile modulus versus weight fraction of C/S composites

Table 4.5 Flexural strength properties

Composites	Break load (N)	Flexural strength (MPa)	Flexural modulus (GPa)
40C/0S	267	64.09	2.14
25C/15S	287.7	69.05	2.73
20C/20S	341.2	81.89	3.06
15C/25S	371.5	89.16	3.40
0C/40S	282.1	67.70	2.64

properties. It was observed that the flexural properties trend is similar to the tensile properties. The loss in strength of single fiber composites can be explained as (1) early failure of the composite upon external loading. (2) weak interfacial bonds between fiber and matrix material leads to the clustering of fibers.

The results presented in Fig. 4.8 show the evolution of flexural modulus according to the fiber weight fractions. It was observed from the results that single fiber Caryota and sisal fiber composites showed a lower flexural modulus than Caryota sisal hybrid composites. Single fiber composites showed 2.14 GPa and 2.64 GPa respectively.

The flexural strength is higher than the tensile strength. Indeed, the dimensions of the test samples are similar for both tests. Only at the center of the composite is stressed in the three-point bending test while the whole is in tension test, then

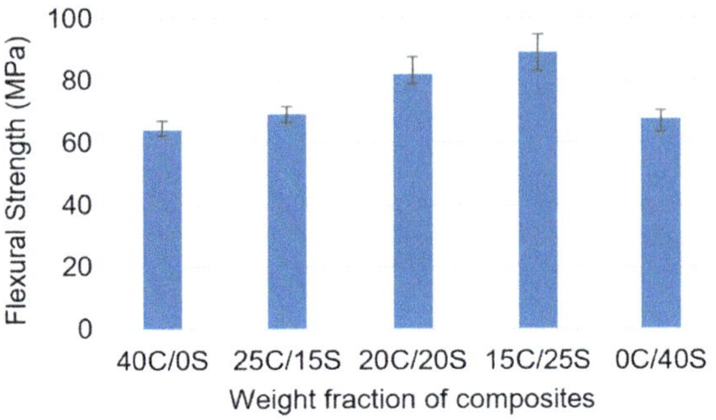

Fig. 4.7 Flexural strength versus weight fraction of C/S composites

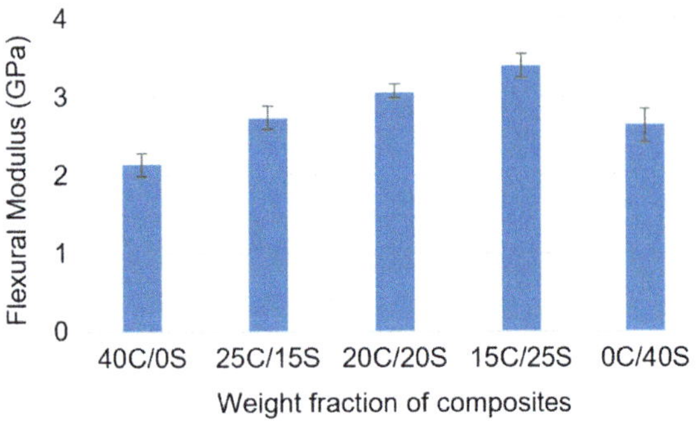

Fig. 4.8 Flexural modulus versus weight fraction of C/S composites

fewer defects are involved in the 3-point bending test. Whereas tangent modulus or tensile modulus is more than the flexural modulus. The reason can be attributed to, in a tensile test, the maximum tensile stresses are experienced throughout the entire volume (and surface area) of composites whereas, in bending test, the maximum tensile stresses are conversely concentrated in a small region on the top surface.

Impact Test. The data presented in Fig. 4.9 was obtained from the impact tests. The specifications of the samples were taken from ASTM standards and the tests were conducted at 21 °C temperature and relative humidity as 28%. It is well known that the impact strength of the composites mainly depends on the properties of reinforcement and matrix material presented in it. It was observed that the hybrid composites showed superior impact strength over single fiber composites. Single sisal fiber composites exhibited the least impact strength as 83.5 J.

Vickers Hardness Test. The hardness values of various weight fractions of composites are given in Fig. 4.10. It was observed from the results, single fiber composites showed the highest hardness values whereas hybrid composites showed the least hardness values. The hybridization impact is not noticed in this case. 20C/20S composite showed the least Vickers hardness values among all composites.

Fig. 4.9 Impact strength versus fiber weight fractions

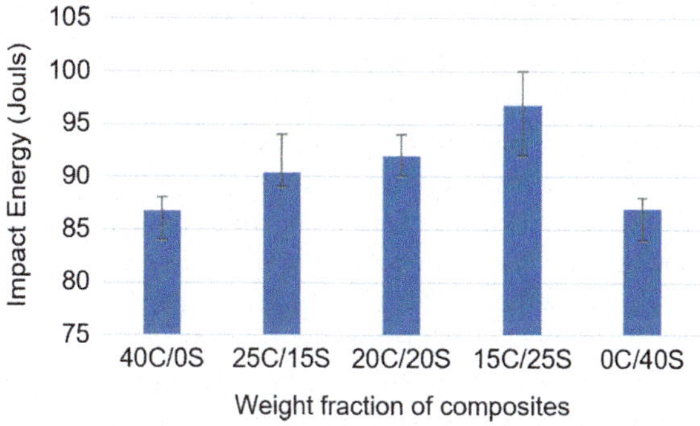

Fig. 4.10 Hardness versus weight fraction of C/S composites

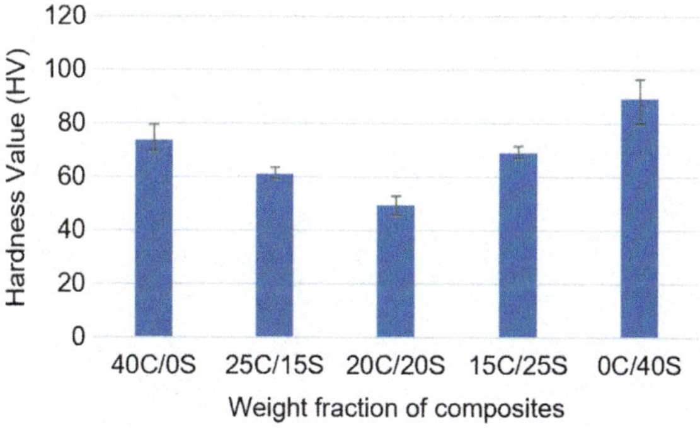

Moisture Analysis. The amount of moisture absorbed by the composites was calculated by using the moisture test. The samples were allowed for the heat treatment process (placed in an oven at 60 °C for 15 min time to eliminate the moisture in it) before placing them in the water container. The sample dimensions were taken from the ASTM standards, and the test was conducted at 21 °C temperature and relative humidity as 28%.

Figure 4.11 shows the % weight gain by the composites with varying weight fractions to time. It was observed that all the composites' samples responded to the moisture (water) when it is exposed. The amount of weight gain by the composites increased with time up to 120 h and remains constant. All hybrid composites showed intermediate results whereas single fiber composites showed the least and highest percentage of weight gain. The reason can be attributed to the presence of void content (porosity) in the composites and the chemical composition (cellulose content) of the fibers. Because the oxygen ($C_6H_{10}O_5$) in the cellulose reacts to hydrogen in the moisture and forms the hydrogen–oxygen bonds with it. Sisal fibers contain more cellulose percentage than Caryota, hence it observed more % of water. The porosity in the composites specimens was observed by using SEM analysis and the results were mentioned in Fig. 4.17. It was observed that hybrid composites have an intermediate percentage of void places compared to single fiber composites. The presence of void places attracts water molecules hence the percentage gain is more.

Scanning Electron Microscopic (SEM) Analysis. The SEM images of the flexural, tensile tests fractured composites and chemical composition at the surface and fractured specimens were taken to analyze the presence of voids, cracks, and the reinforcement and matrix adhesion (bonding) behavior.

The above Fig. 4.12 shows the micrographic spectrum acquisition for single sisal fiber composite to analyze the chemical composition in it.

Based on the spectrum analysis of a specimen, single sisal fiber hybrid composites the presence of composition on the surface is shown in Fig. 4.13. The data presented in the above graph is shown in the following Table 4.6. The structure is expressed by the atomic weight and number of each element presented in it. The weight % of carbon is more than the weight % of oxygen, this reflects that the matrix concentration is more at the surface of the composite than reinforcement. Due to this smooth surface finish will occur to the composite.

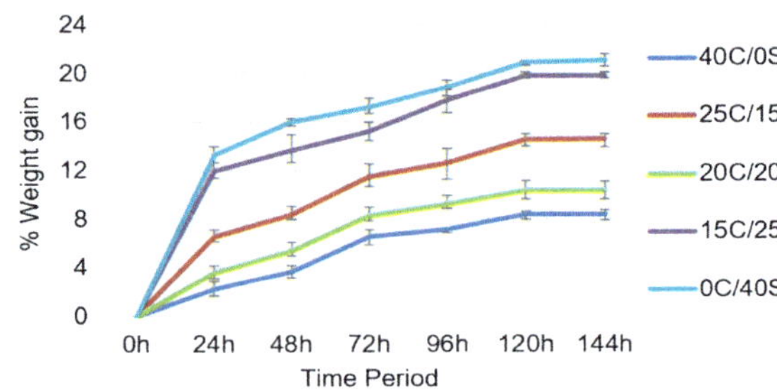

Fig. 4.11 Weight gain in % by the composites vs. time period

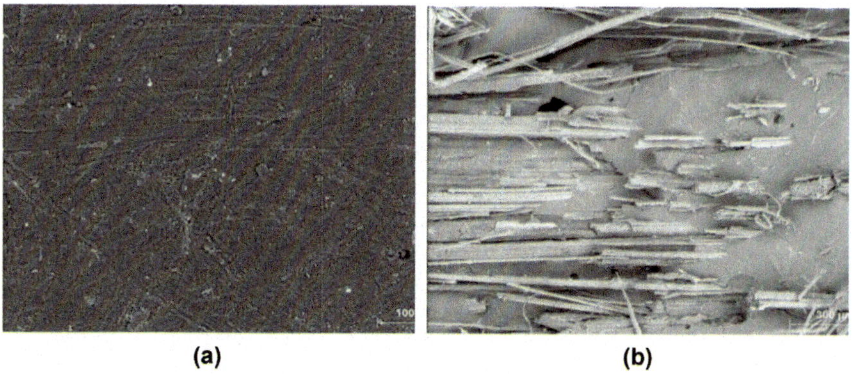

Fig. 4.12 SEM micrographs of single sisal fiber composite (0C/40S) **a** surface and **b** fractured position

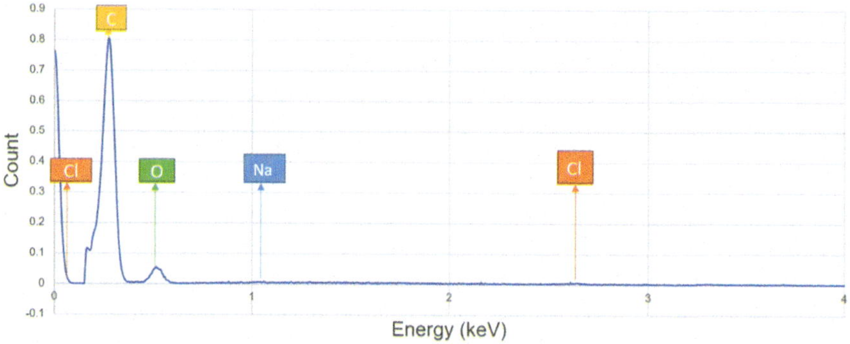

Fig. 4.13 Spectrum analysis of the chemical structure of single sisal fiber hybrid composite surface position obtained at an intensity of 4 keV

Table 4.6 Chemical structure of single sisal fiber hybrid composite at surface position

Element symbol	Element name	Series name	Atomic weight	Weight Conc. (%)	Atomic Conc. (%)
C	Carbon	K-series	12	68.68	74.66
O	Oxygen	K-series	16	30.75	25.09
Na	Sodium	K-series	23	0.17	0.10
Cl	Chlorine	K-series	35.5	0.40	0.15

Based on the spectrum analysis of a specimen, single sisal fiber hybrid composites the presence of composition on the surface is shown in Fig. 4.14. The data presented in the above graph is shown in the following Table 4.7. The structure is expressed by the atomic weight and number of each element presented in it. It was observed that the weight % of carbon is nearly equal to the weight % of oxygen, this reflects

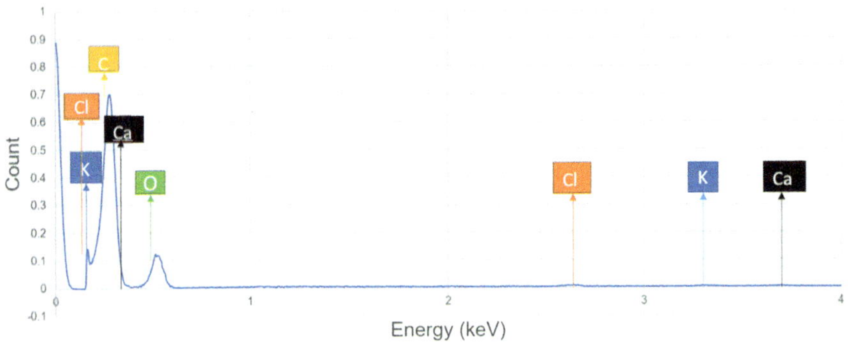

Fig. 4.14 Spectrum analysis of the chemical structure of single sisal fiber hybrid composite fractured position obtained at an intensity of 4 keV

that the matrix concentration and reinforcements are in proper proportion inside the composite. This reflects strong bonding between matrix and reinforcement in the composite and also, improvement in the strength of the composite.

After the analysis of chemical concentration presented at the surface and fractured position of single sisal fiber hybrid composite, we can conclude that the percentage of carbon is increased in the surface compared to the fractured (inside) position which denotes that epoxy concentration is more at the surface. Whereas oxygen percentage more at the fractured position which denotes that fiber concentration is more at the breakpoint.

Figure 4.15 shows the fracture surface of the flexural test composites as a function of hybridization. A brittle fracture of reinforcement and matrix material was observed from the results. The reason can be attributed to the strong interfacial bonds between fibers and epoxy resin. It can be seen from the above figures that the presence of voids and cracks leads to weakening the strength of the composites. It was observed that single fiber composites (Fig. 4.15a, e) have more cracks and air gaps in them when compared to the hybrid composites, this is due to lack of fiber/resin adhesion and poor bonding nature.

The fracture properties of composites after the tensile test were inspected through SEM and it is given in Fig. 4.16. It is observed that the cluster of matrix shards cling to

Table 4.7 Chemical structure of single sisal fiber hybrid composite at fractured position

Element symbol	Element name	Series name	Atomic weight	Weight Conc. (%)	Atomic Conc. (%)
C	Carbon	K-series	12	52.56	59.98
O	Oxygen	K-series	16	46.18	39.57
Cl	Chlorine	K-series	35.5	0.46	0.18
K	Potassium	K-series	39.09	0.38	0.13
Ca	Calcium	K-series	40.08	0.42	0.14

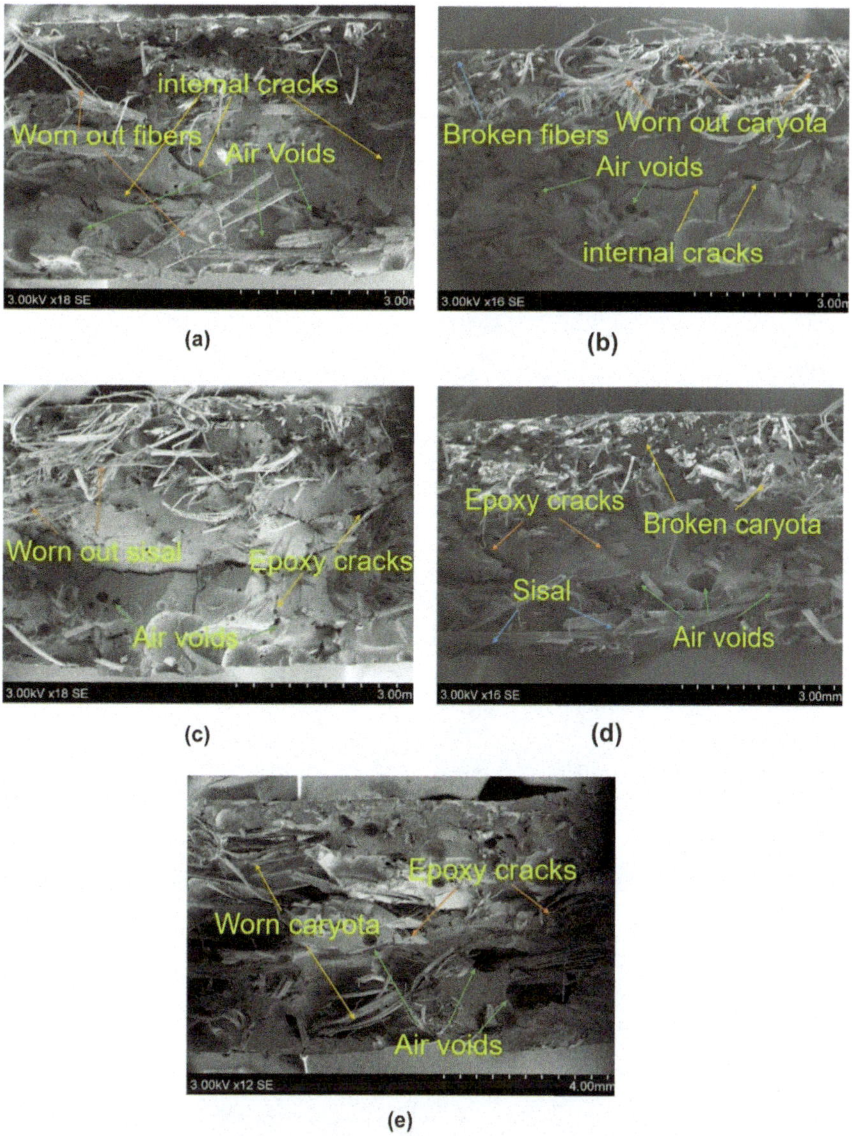

Fig. 4.15 SEM images for fractured composites of flexure test specimens **a** 0C/40S, **b** 15C/25S, **c** 20C/20S, **d** 25C/15S, **e** 40C/0S

the fiber reinforcements and enclosed it. This helps the composite to worn out when the external load acting on it when it reaches its maximum strength. The fiber pulled out needs extra strength compared to fiber fracture. It can be seen that in the case of single fiber composites (Fig. 16a,e) the fibers are separated from the matrix due to poor adhesion. Wherein the case of hybrid composites some of the matrix material is presented at the top of fibers. This indicates the strong crosslinking reaction between reinforcement and matrix materials. That gives extra support to the composites to worn out. Moreover, it was observed that the presence of cracks and void places are less in hybrid composites compared to the single fiber composites.

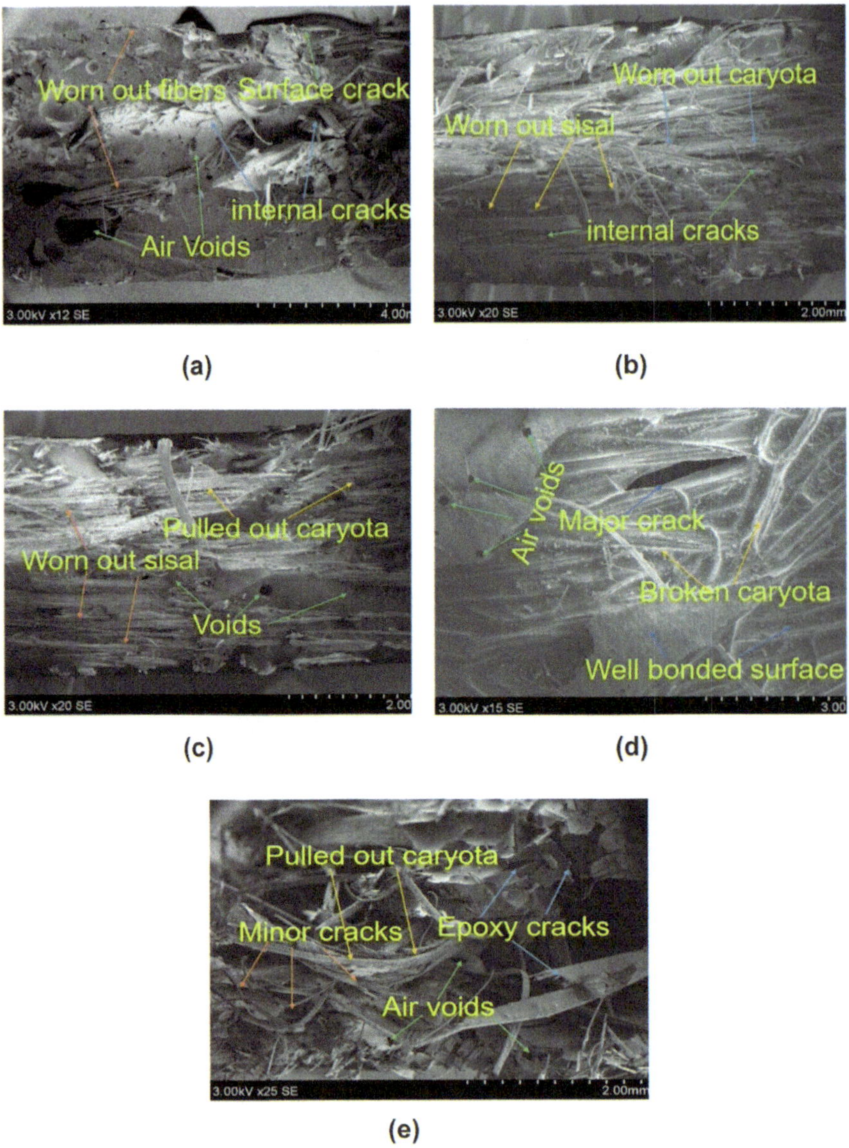

Fig. 4.16 SEM images for fractured composites of tensile test specimens **a** 0C/40S, **b** 15C/25S, **c** 20C/20S, **d** 25C/15S, **e** 40C/0S

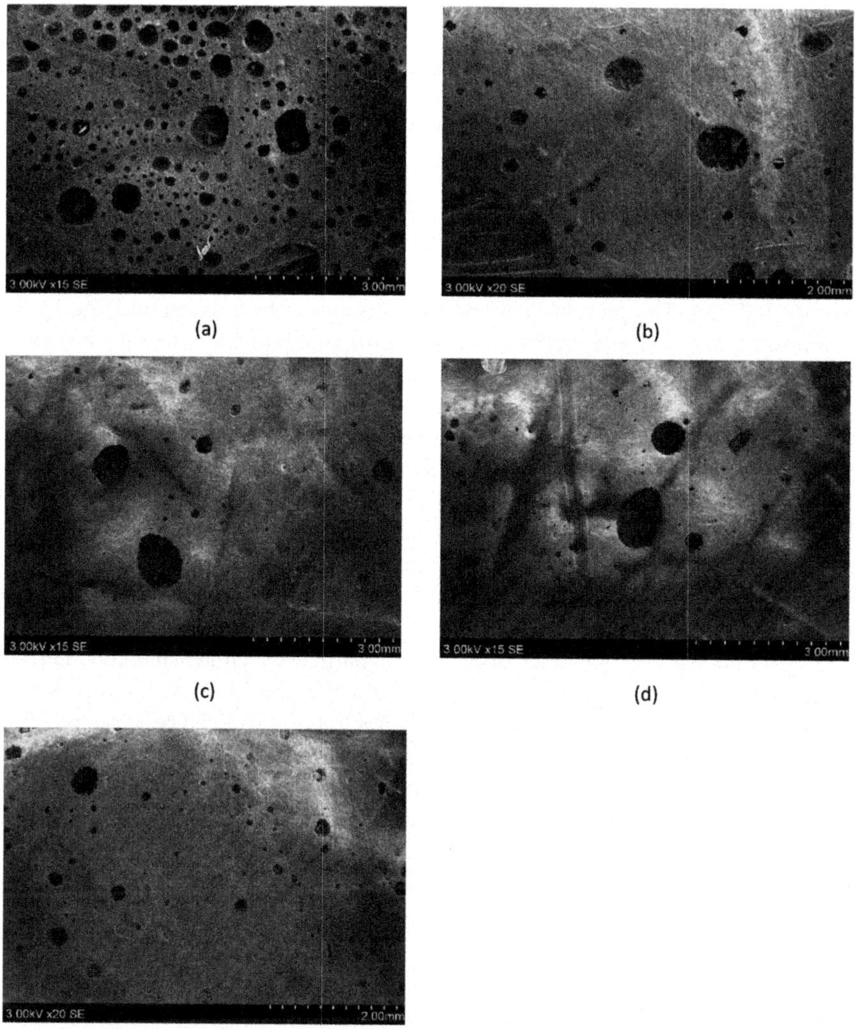

Fig. 4.17 SEM images for porosity (void content) presented in the fabricated composites **a** 0C/40S, **b** 15C/25S, **c** 20C/20S, **d** 25C/15S, **e** 40C/0S

The presence of air voids and gaps in the composite surfaces may lead to weakening the strength of the composite and also the composite may absorb more water when it faces the wet medium. Figure 4.17 shows the presence of porosity content in all the fabricated composites, it was observed that 0C/40S composite has the more void percentage whereas 40C/0S composite has the least percentage of void content. All the hybrid composites have intermediate percentages and close to each other. The reasons can attribute as (1) strong internal bonding between reinforcement and matrix in the hybrid composites. (2) fabrication errors. (3) atmospheric conditions such as temperature and humidity.

4.4 Discussion

Previous research was concentrated on the fabrication of banana, pineapple, hemp, flax, and sisal fibers-based hybrid composites for lightweight and biomedical applications [43–49]. The present research was concentrated on developing the hybrid composites with Caryota (derived from fishtail palm plants) and sisal fiber-based hybrid composites with varying weight fractions.

The Caryota and sisal fiber-reinforced epoxy resin-based hybrid composites were fabricated by using the hand layup technique. The developed hybrid composites have shown superior mechanical properties. In the previous research, banana/pineapple-based hybrid composites showed flexural properties ranging from 37 to 53 MPa and hemp/flax hybrid composites showed that in between 45 and 63 MPa whereas Caryota and sisal-based hybrid composites showed in between 65 and 89 MPa which is far higher than other hybrid composites. Similarly, there is a considerable improvement in the tensile properties as well.

For future aspects, implementation of nanoparticles such as nano clay. Ferrous oxide and silver particles in bio-based epoxy resin foreseen. Moreover, the orientation of fiber materials in the composite materials has to be considered because, previous researchers proved that, the orientation of reinforcement may increase the torsional strength and also crosslinking between matrix and fiber material. The bio-based composites could be used for biomedical and microfluidic applications.

4.5 Conclusions

Caryota and sisal fiber-based epoxy resin hybrid composites were developed by using the hand layup technique and the mechanical properties were investigated. Experimental results stated that hybrid composites showed superior properties. From the tensile test, it was found that 15C/25S hybrid composites have shown 38 MPa whereas single fiber composites (0C/40S, 40C/0S) showed 25 and 22 MPa respectively. From flexural properties, it was found that the hybrid composites showed improved flexural properties than single fiber composites. The 15C/25S hybrid composites showed flexural strength and flexural modulus as 89.16 MPa and 3.40 GPa. Single Caryota fiber composites showed the least strength and modulus as 64.09 MPa and 2.14 GPa. Coming to the impact strength 15C/25S composites showed as 97 J and in single sisal fiber composites, it was observed that 83.5 J. in all the tests hybridization effect was observed clearly except the hardness test. In the hardness test, single fiber composites showed superior properties to hybrid composites. The reason could be fabrication errors or poor bonding between matrix and reinforcement material.

Thus, SEM analysis showed that hybrid composites have fewer defects such as voids, minor and major cracks, poor adhesion between matrix and reinforcement material than single fiber composites.

Moisture absorption measurements stated that all the composites showed a constant increase up to 120 h and remain constant. Hybrid composites showed intermediate results whereas single fiber composites showed the highest and lowest rate of absorption. By considering all the results Caryota and sisal fiber hybrid composites showed superior mechanical properties than single fiber composites. And Caryota and sisal fibers are potential alternatives in fiber composite fabrication.

References

1. Atmakuri A, Palevicius A, Kolli L, Vilkauskas A, Janusas G (2021) Development and analysis of mechanical properties of Caryota and sisal natural fibers reinforced epoxy hybrid composites. Polymers 13(6):1–19. Basel: MDPI. ISSN 2073-4360. Art no. 864. https://doi.org/10.3390/polym13060864.
2. Sanjay MR, Siengchin S, Parameswaranpillai J, Jawaid M, Pruncu CI, Khan A (2019) A comprehensive review of techniques for natural fibers as reinforcement in composites: preparation, processing and characterization. Carbohydr Polym 207:108–121
3. Barbosa AQ, da Silva LFM, Abenojar J, del Real JC, Paiva RMM, Öchsner A (2015) Kinetic analysis and characterization of an epoxy/cork adhesive. Thermochim Acta 604:52–60
4. He H, Li K, Wang J, Sun G, Li Y, Wang J (2011) Study on thermal and mechanical properties of nano-calcium carbonate/epoxy composites. Mater Des 32:4521–4527
5. Jin FL, Park SJ (2009) Thermal stability of trifunctional epoxy resins modified with nanosized calcium carbonate. Bull Korean Chem Soc 30:334–338
6. Mishra V, Desai J, Patel KI (2017) High-performance waterborne UV-curable polyurethane dispersion based on thiol–acrylate/thiol–epoxy hybrid networks. J Coat Technol Res 14:1069–1081
7. Chowaniec A, Sadowski L, Żak A (2020) The chemical and microstructural analysis of the adhesive properties of epoxy resin coatings modified using waste glass powder. Appl Surf Sci 504:1–13
8. Bard S, Demleitner M, Weber R, Zeiler R, Altstädt V (2019) Effect of curing agent on the compressive behavior at elevated test temperature of carbon fiber-reinforced epoxy composites. Polymers 11:943
9. Garcia FG, Leyva ME, Oliveira MG, de Queiroz AAA, SimõEs AZ (2010) Influence of chemical structure of hardener on mechanical and adhesive properties of epoxy polymers. J Appl Polym Sci 117:2213–2219
10. Deng B, Guo Y, Song S, Sun S, Zhang H (2016) Effect of core-shell particles dispersed morphology on the toughening behavior of PBT/PC blends. J Polym Res 23:1–8
11. Chen J, Chen T, Liu XP, Xiao LR, Qian QR, Xu J, Huang BQ, Chen QH (2014) Effect of epoxy-functionalized acrylonitrile-butadiene-styrene on the morphology and rheological properties of poly (butylenes terephthalate)/poly(acrylonitrile-styrene-acrylate) blends. Adv Mater Res 893:228–234
12. Fu N, Li G, Zhang Q, Wang N, Qu X (2014) Preparation of a functionalized core-shell structured polymer by seeded emulsion polymerization and investigation on toughening poly (butylene terephthalate). RSC Adv 4:1067–1073
13. Wu G, Ji W, Sun S, Zhang H (2008) Effect of epoxy-functionalised core-shell particles on properties of poly (butylenes terephthalate) (PBT). Polym Polym Compos 16:271–276

14. Hao Y, Li W, Sun S, Zhang H, Gao G, Dong L (2015) Poly(butylene terephthalate) toughening with butadiene-epoxy-functionalized methyl methacrylate core-shell copolymer. J Macromol Sci Part B Phys 54:1267–1281
15. Sharma AK, Bhandari R, Aherwar A, Rimašauskiene R (2020) Matrix materials used in composites: a comprehensive study. Mater Today Proc 21:1559–1562
16. Kumar A, Singh RC, Chaudhary R (2020) Recent progress in production of metal matrix composites by stir casting process: an overview. Mater Today Proc 21:1453–1457
17. Shahzad A (2013) A study in physical and mechanical properties of hemp fibres. Adv Mater Sci Eng 2013:1–9
18. Jawaid M, Abdul Khalil HPS (2011) Cellulosic/synthetic fibre reinforced polymer hybrid composites: a review. Carbohydr Polym 86:1–18
19. Ramamoorthy SK, Skrifvars M, Persson A (2015) A review of natural fibers used in biocomposites: plant, animal and regenerated cellulose fibers. Polym Rev 55:107–162
20. Chan ML, Lau KT, Wong TT, Ho MP, Hui D (2011) Mechanism of reinforcement in a nanoclay/polymer composite. Compos Part B Eng 42:1708–1712
21. Velmurugan R, Manikandan V (2007) Mechanical properties of palmyra/glass fiber hybrid composites. Compos Part A: Appl Sci Manuf 38:2216–2226
22. Pappu A, Thakur VK (2017) Towards sustainable micro and nano composites from fly ash and natural fibers for multifunctional applications. Vacuum 146:375–385
23. Nayak S, Kumar Khuntia S (2019) Development and study of properties of Moringa oleifera fruit fibers/polyethylene terephthalate composites for packaging applications. Compos Commun 15:113–119
24. Wang X, Xie J, Zhang H, Zhang W, An S, Chen S, Luo C (2019) Determining the lignin distribution in plant fiber cell walls based on chemical and biological methods. Cellulose 26:4241–4252
25. Jacob M, Thomas S, Varughese KT (2004) Natural rubber composites reinforced with sisal/oil palm hybrid fibers: tensile and cure characteristics. J Appl Polym Sci 93:2305–2312
26. Sreekala MS, George J, Kumaran MG, Thomas S (2002) The mechanical performance of hybrid phenol-formaldehyde-based composites reinforced with glass and oil palm fibres. Compos Sci Technol 62:339–353
27. Swolfs Y, Crauwels L, Van Breda E, Gorbatikh L, Hine P, Ward I, Verpoest I (2014) Tensile behaviour of intralayer hybrid composites of carbon fibre and self-reinforced polypropylene. Compos Part A Appl Sci Manuf 59:78–84
28. Abou El-Mal HSS, Sherbini AS, Sallam HEM (2015) Mode II fracture toughness of hybrid FRCs. Int J Concr Struct Mater (9):475–486
29. Shubaili MA, Sallam HEM (2016) Discussion: Mechanical properties of hybrid fiber-reinforced concrete-analytical modelling and experimental behaviour. Mag Concr Res 68:1183–1186
30. Al-Tersawy SH, El-Sadany RA, Sallam HEM (2021) Long term behavior of normal weight concrete containing hybrid nanoparticles subjected to gamma radiation. Arch Civ Mech Eng 21:1–18
31. Alamri H, Low IM (2012) Mechanical properties and water absorption behaviour of recycled cellulose fibre reinforced epoxy composites. Polym Test 31:620–628
32. Alhazmi WH, Jazaa Y, Mousa S, Abd-Elhady AA, Sallam HEM (2021) Tribological and mechanical properties of epoxy reinforced by hybrid nanoparticles. Lat Am J Solids Struct 18:1–16 in press
33. Colomer-Romero V, Rogiest D, García-Manrique JA, Crespo JE (2020) Comparison of mechanical properties of hemp-fibre biocomposites fabricated with biobased and regular epoxy resins. Materials 13:5720
34. Saba N, Abdan PK, Ibrahim NA (2016) Dynamic mechanical properties of oil palm nano filler/kenaf/epoxy hybrid nanocomposites. Constr Build Mater 124:133–138
35. Jawaid M, Alothman OY, Paridah MT, Khalil HPSA (2014) Effect of Oil Palm and jute fiber treatment on mechanical performance of epoxy hybrid composites. Int J Polym Anal Charact 19:62–69

36. Hanan F, Jawaid M, Md Tahir P (2020) Mechanical performance of oil palm/kenaf fiber-reinforced epoxy-based bilayer hybrid composites. J Nat Fibers 17:155–167
37. Cordeiro EP, Pita VJRR, Soares BG (2017) Epoxy-fiber of Peach Palm trees composites: the effect of composition and fiber modification on mechanical and dynamic mechanical properties. J Polym Environ 25:913–924
38. Hanan F, Jawaid M, Paridah MT, Naveen J (2020) Characterization of hybrid oil palm empty fruit bunch/woven kenaf fabric-reinforced epoxy composites. Polymers 12:2052
39. Ramesh M, Palanikumar K, Reddy KH (2013) Comparative evaluation on properties of hybrid glass fiber-sisal/jute reinforced epoxy composites. Procedia Eng. 51:745–750
40. Mochane MJ, Mokhena TC, Mokhothu TH, Mtibe A, Sadiku ER, Ray SS, Ibrahim ID, Daramola OO (2019) Recent progress on natural fiber hybrid composites for advanced applications: a review. Express Polym Lett 13:159–198
41. Gholampour A, Ozbakkaloglu T (2020) A review of natural fiber composites: properties, modification and processing techniques, characterization, applications. J Mater Sci 55:829–892
42. Shivamurthy B, Murthy K, Joseph PC, Rishi K, Bhat KU, Anandhan S (2015) Mechanical properties and sliding wear behavior of Jatropha seed cake waste/epoxy composites. J Mater Cycles Waste Manag 17:144–156
43. Cavalcanti DKK, Banea MD, Neto JSS, Lima RAA, da Silva LFM, Carbas RJC (2019) Mechanical characterization of intralaminar natural fibre-reinforced hybrid composites. Compos Part B Eng 175:1–8
44. Atmakuri A, Palevicius A, Griskevicius P, Janušas G (2019) Investigation of mechanical properties of hemp and flax fibers hybrid composites for biomedical applications. Mechanika 25:149–155
45. Pereira AL, Banea MD, Neto JSS, Cavalcanti DKK (2020) Mechanical and thermal characterization of natural intralaminar hybrid composites based on sisal. Polymers 12:866
46. Bazan P, Mierzwiński D, Bogucki R, Kuciel S (2020) Bio-based polyethylene composites with natural fiber: mechanical, thermal, and ageing properties. Materials 13:2595
47. Singh S, Deepak D, Aggarwal L, Gupta VK (2014) Tensile and flexural behavior of hemp fiber reinforced virgin-recycled HDPE matrix composites. Procedia Mater Sci 6:1692–1702
48. Atmakuri A, Palevicius A, Siddabathula M, Vilkauskas A, Janusas G (2020) Analysis of mechanical and wettability properties of natural fiber-reinforced epoxy hybrid composites. Polymers 12:2827
49. Giridharan R (2019) Preparation and property evaluation of Glass/Ramie fibers reinforced epoxy hybrid composites. Compos Part B Eng 167:342–345

Chapter 5
Numerical and Experimental Analysis of Natural-Fiber-Reinforced Hybrid Polymer Composites

Abstract The effect of matrix material on the mechanical properties of natural fiber-reinforced hybrid composites was examined through a comparison of their experimental and numerical analysis results. In this chapter, hemp and flax fibers were utilized as reinforcement, while epoxy resin and ecopoxy resin, combined with a hardener, were employed as matrix materials (Atmakuri et al. in Polym Basel MDPI. 14:1–18, 2022 [1]). To investigate the influence of the matrix material, two sets of hybrid composites were synthesized by altering the matrix material. The composite samples were fabricated using the compression-molding technique followed by a hand layup process. A total of five distinct composites were produced by varying the weight fraction of fiber material in each set according to the rule of the hybridization process. Following fabrication, the mechanical properties of the composite samples were evaluated, and morphological studies were conducted using SEM–EDX analysis. The flexural-test fractured specimens were examined using a scanning electron microscope (SEM). Furthermore, theoretical analysis of the elastic properties of hybrid composites was performed using the Halpin–Tsai approach. The findings revealed that the hybrid composites exhibited superior properties compared to individual fiber composites. Overall, epoxy resin matrix composites demonstrated superior properties in comparison to ecopoxy matrix composites.

5.1 Introduction

Natural fiber composites have garnered considerable attention in recent years owing to the growing concerns regarding the environmental impact of synthetic fiber composites [2]. The incorporation of natural fibers within thermosetting or thermoplastic polymer matrices has sparked significant interest among researchers, attributable to their cost-effectiveness, ease of availability, environmental compatibility, and their competitive edge over synthetic composites [3, 4]. Natural-fiber-reinforced polymer composites have found applications across diverse industries,

A. Atmakuri et al., *Fiber-Based Hybrid Composites for Engineering*, Springer Series on Polymer and Composite Materials, https://doi.org/10.1007/978-3-031-93144-4_5

such as in automotive parts, aerospace components, sporting goods, marine applications, military supplies, and within the biomedical, microfluidics, and construction sectors [5–7]. Their reduced self-weight, heightened specific strength, inherent formability, and robust corrosion and fatigue resistance further contribute to the increasing prominence of natural-fiber-reinforced composites. Conversely, the application of natural composites has been constrained due to their inferior mechanical and thermo-physical properties in comparison to synthetic composites and conventional structural materials [8, 9]. Plant-derived natural fibers such as banana, pineapple, abaca, jute, okra, sisal, hemp, flax, caryota, kenaf, and date palm fibers have demonstrated substantial potential as reinforcement in thermosetting and thermoplastic polymer composites and are being utilized across a broad spectrum of applications [10, 11].

In composite materials, both the reinforcement and matrix constituents are integral to their functionality. The reinforcement component functions as the primary load-bearer, while the matrix component serves as the distributor of the applied load within the composite structure [12, 13]. Matrix materials are categorized into thermoset and thermoplastic polymers. Thermoplastic materials are amenable to reuse following the curing process, provided they undergo appropriate heat treatment. Examples of thermoplastic polymers include polyester, polypropylene, nylon, and Teflon. In contrast to thermoplastic polymers, thermoset polymers are not reusable post-curing. Once the liquid thermoset resin is condensed (solidified) at the requisite temperature, it becomes non-recyclable. Typical thermoset resins comprise epoxies, polyamides, and vinyl esters. Over recent years, the application of thermoset polymers has gained more prominence compared to thermoplastic polymers, owing to their extensive applicability across diverse sectors [14–16]. Among polymer composites, those based on epoxy resin thermoset polymers have garnered significant attention due to their excellent adhesive qualities, minimal shrinkage, rapid curing (transition from liquid to solid state), and user-friendly nature [17, 18].

The static and dynamic mechanical attributes of polymer composites reinforced with natural fibers predominantly depend upon the quality and quantity of the fiber materials [19, 20], in addition to the interfacial strength and adhesion characteristics between the components of the polymer composite [21]. The hybridization of composite materials presents an alternative method to enhance their mechanical performance. Hybridization is defined as the production of a composite through the amalgamation of two or more reinforcements under an identical matrix material, resulting in what are termed hybrid composites [22, 23]. Numerous researchers have examined the mechanical properties of hybrid composites and demonstrated advancements in their characteristics. The primary benefit of hybridization is that the integration of reinforcements can be achieved through various methodologies, including blended continuous fibers, short fibers, discontinuous fibers, and particulate-matter fibers [24–26]. A significant majority of hybrid compositions emphasize natural fibers combined with synthetic fibers such as glass and carbon fibers, offering a range of properties superior to those of individual fiber composites. Nonetheless, natural fiber hybrid composites have gained increased popularity over synthetic fiber composites due to environmental considerations [27].

The chemical composition of natural fibers is crucial in plant-based fiber composites; however, environmental conditions such as climatic variations and geographical differences influence the composition of natural fibers [28]. As a result, their composition is often expressed in terms of maximum and minimum values [29]. Typically, plant-based fibers are primarily composed of cellulose, hemicellulose, and lignin. The cellulose content and aspect ratio of natural fibers are significant determinants of their mechanical performance [30, 31]. In contrast, the mechanical properties of artificial (synthetic) fibers, which lack these chemical compositions, remain relatively consistent. Natural fibers, in turn, develop in open environments with the influence of air, light, water, soil, among others, causing variations over time, which in turn affect fiber properties [32, 33]. Certain research studies indicate that fiber composites reinforced with cellulose-rich plant-based natural fibers like hemp, flax, and banana exhibit superior properties compared to composites reinforced with natural fibers containing lower cellulose content [34]. Consequently, in the present study, hemp and flax fibers were employed as reinforcements to examine the impact of matrix material on mechanical performance.

In recent times, a significant number of researchers have expressed interest in exploring natural fiber composites in conjunction with thermoset polymers. Despite the numerous advantages associated with natural fiber composites, their utilization remains constrained by issues related to moisture absorption and flame-retardancy characteristics. Pan and Zhong [35] developed a micromechanical model to elucidate the impact of moisture on the degradation of composite materials. The mechanical degradation and swelling properties of natural fiber composites were examined using the Mori–Tanaka model, incorporating various shapes. The findings indicated that an appropriate choice of matrix material can prevent mechanical degradation and reduce moisture absorption by the natural fiber composite. Neves et al. [36] examined the mechanical properties of hemp fiber composites with different matrix materials, comparing epoxy and polyester resins. Their investigation revealed that hemp fiber composites utilizing epoxy resin exhibit superior properties compared to those using polyester resin and suggested these composites have potential applications in armed contexts. Sapuan et al. [37] researched the mechanical properties of woven banana-fiber-reinforced epoxy composites. They developed a theoretical model based on the ANOVA technique to compare theoretical and experimental results, observing slight variations between the two. Several researchers have also provided useful and comprehensive review articles on natural-fiber-reinforced hybrid composites, underlining their significance in terms of mechanical performance [10, 38, 39]. Summerscales et al. [40] conducted a comprehensive review of various statistical models pertinent to natural fiber composites. This study also delineates the different harvesting techniques and cultivation conditions of natural fibers. Alhazmi et al. [41] explored the tribological and mechanical properties of hybrid composites incorporating epoxy resin. The findings indicated that the addition of nanofiller materials did not notably enhance the wear resistance of the composite material. Singh et al. [42] examined the mechanical properties of hemp fiber composites. To assess these properties, tensile and flexural tests were performed, revealing an increase in mechanical properties with increasing fiber content, reaching a peak at 40%, after which a decline was observed. Moudood et al. [43] investigated the mechanical properties of flax fibers under diverse environmental conditions, with experimental measurements aligning closely with the statistical outcomes.

Zhong et al. [44] conducted a review of the challenges encountered in the numerical simulations and modeling of natural fiber composites. The microstructures of these composites were elaborated upon by highlighting their properties, which necessitate thorough examination prior to presentation. The authors also engaged in a discussion and synthesis of prior numerical studies. In contrast to artificial fibers, natural fiber composites notably excel in the realm of numerical methodologies [45]. The existing specific challenges within this field are recapitulated to provide potential avenues for future exploration. Jose da Silva et al. [46] undertook a numerical and experimental investigation of natural-fiber composites. This study chiefly concentrated on sisal and banana natural-fiber-reinforced epoxy-resin polymer composites. Tensile tests were administered to substantiate the numerical findings. To corroborate the experimental results, both three-dimensional and two-dimensional models were devised with regard to the elastic properties of each constituent of the composite material. The two-dimensional analysis was grounded in the rule of mixtures (ROM), while the three-dimensional analysis was executed using ANSYS software. The outcomes indicated that variations in Poisson's ratio values did not exert an influence on the finite element analysis results. Sathishkumar et al. [47] conducted an investigation into the experimental and numerical analysis of filler material on sisal-fiber-reinforced polymer composites. The examination of mechanical properties was performed utilizing flexural and tensile testing. The two-dimensional analysis of polymer composites employed the Rule of Mixtures method along with the Halpin–Tsai approach. The three-dimensional analysis was executed via the ANSYS workbench, taking into account the elastic properties of the materials utilized for fabrication. The findings demonstrated an enhancement in mechanical properties attributable to the inclusion of filler material within the composite materials. Prasanthi et al. [48] examined the influence of waviness on sisal fiber composites through numerical and experimental methodologies. The objective of the study was to comprehend the impact of waviness on fiber composites and to compare the results using both analytical and experimental approaches. The analytical investigations were conducted utilizing the ANSYS model, and the findings indicated that waviness exerts a significant impact on fiber composites. Balasubramanian et al. [49] conducted an analysis of the mechanical properties of natural fibers through both experimental and finite element analysis (FEA) methodologies. Hybrid composites were fabricated employing flax, sisal, and aloe vera fibers as reinforcements, with epoxy resin serving as the matrix material. The mechanical properties were evaluated via tensile, flexural, and impact tests. The finite element analysis was employed to ascertain the stress, deformation, and displacement of the composite materials, after which the results were compared.

From the literature reviewed, two notable insights are discerned. Firstly, the mechanical properties of natural fibers can be enhanced through a hybridization process. Secondly, it is possible to conduct numerical and theoretical analyses of natural fibers by assuming fixed properties utilizing finite element analysis and micromechanics. Based on this knowledge, the primary objective of our research was to investigate the influence of matrix materials on the mechanical properties of natural fiber hybrid composites. The focus was on understanding the effect of matrix material on the mechanical properties of natural-fiber-reinforced epoxy and ecopoxy hybrid composites. In this study, hemp and flax fibers served as reinforcements, while epoxy and ecopoxy materials were used as matrix materials to provide a basis for comparison. The findings were evaluated using both experimental and numerical methods. The mechanical properties of the hybrid composites were assessed through flexural and interlaminar shear properties. Furthermore, wettability properties were examined using contact angle measurement, and the morphological investigations were conducted with SEM–EDX spectrum analysis. The theoretical analysis of composite materials was undertaken using the Halpin–Tsai method, addressing the effect of matrix material on the mechanical properties of natural fiber hybrid composites.

5.2 Materials and Methods

Materials. For the purposes of conducting the present research, hemp and flax fibers were sourced from Naturalus Pluostas UAB, located in Vilnius, Lithuania. The epoxy resin and its corresponding hardener were obtained from Composite 24 in Riga, Latvia, while the ecopoxy resin, which comprises 36% bioactivity, along with its hardener were acquired from EcoPoxy EU in Warsaw, Poland. Teak wood was utilized for the preparation of the molding material.

Preparation of Reinforcement and Matrix Materials. The reinforcement materials were initially provided in bundled form. Each fiber was meticulously selected by removing undesirable components, such as thick stems and small pellets, through a manual sifting process. Upon separation of waste materials, the fibers underwent water treatment followed by a drying process until all moisture was eliminated. Nonetheless, the fibers were unsuitable for fabrication due to their smooth surface finish and insufficient adhesive properties. Consequently, the separated fibers underwent a chemical treatment process. They were treated with a 5% NaOH solution and subsequently rinsed with distilled water. After alkaline treatment, the fibers were oven-dried in preparation for the fabrication process. The matrix materials employed for the fabrication process comprised epoxy and ecopoxy resin along with a hardener material. The proportion of the matrices was maintained at a 2:1 ratio, in accordance with the material specifications provided by the supplier. The required quantities of resin and hardener were placed in a plastic container and stirred with a plastic stirrer for four minutes. The container walls were thoroughly scraped to ensure a proper mix, and the solution was allowed to settle for two minutes for clarity. This procedure was repeated for both matrix materials. Once the mixed solution had settled, it was used in the fabrication process.

Composite Sample Preparation. The weight fractions of the reinforcement and matrix materials were considered as per the rule of the hybridization process. This states that a composite sample is fabricated by using two or more fiber laminates under the same resin material with a 0.4 Wf.% [50, 51]. In the present research work, hemp (H) and flax (F) fiber hybrid composites were fabricated in the presence of epoxy resin and ecopoxy resin. The composites were fabricated by using the compression molding technique followed by the hand-layup technique. The required weight proportions of fibers were placed uniformly in the mold material and compressed to acquire the mold shape and then the epoxy resin was applied to it. The fibers were placed in uniform direction ("0" degrees) for neat hemp and flax composites whereas in the hybrid case the fibers were placed in uniform-direction layer-by-layer mode and for testing the orientation of fibers was not considered. The schematic representation of the ply sequence for ecopoxy composites is given in the Fig. 5.1 and a similar sequence is followed for the epoxy-matrix-based hybrid composites. The mold material was covered with a backing paper and then compressed with a uniform load for 72 h. Once the curing process was completed the composites were subjected to the heat treatment process. A total of five different weight fractions of composites were fabricated (40H/0F, 25H/15F, 20H/20F, 15H/25F, and 0H/40F). In all cases the fibers were laid down horizontally with a 0° angle. The sample weight fractions are given in Table 5.1. All the samples were fabricated as per ASTM testing standards.

Flexural Test. To measure the flexural properties such as flexural strength and modulus of the composite samples, a universal testing machine (Tinus Olsen H10K) was used. To test the samples, the 3-point bending test was preferred over the 4-point bending test because the external load acts at the center of the composite so load distribution was uniform [52]. The test was conducted at a constant crosshead speed of 10 mm/min and the strain rate was 0.10 mm/min for all specimens. All

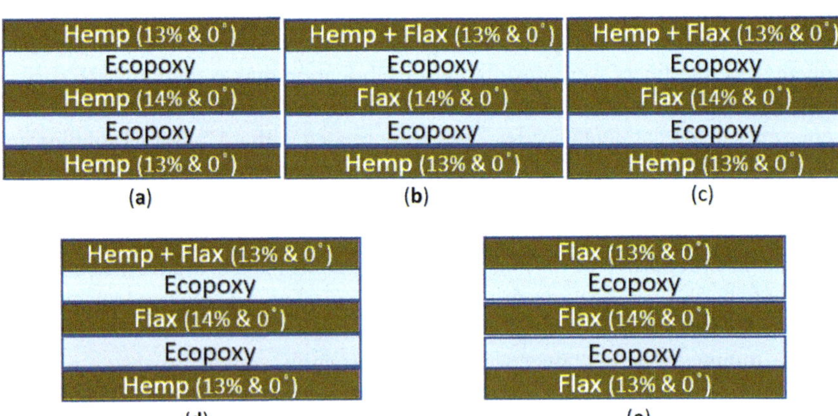

Fig. 5.1 Schematic layout of the ply sequence for ecopoxy-matrix-based hybrid composites. **a** 40H/0F, **b** 25H/15F, **c** 20H/20F, **d** 15H/25F, and **e** 0H/40F

Table 5.1 The weight fraction of composite samples for epoxy/ecopoxy matrices

Composites	Hemp (%)	Flax (%)	Total reinforcement Vol. (%)	Total matrix Vol. (%)
40H/0F	40	0	40	60
25H/15F	25	15	40	60
20H/20F	20	20	40	60
15H/25F	15	25	40	60
0H/40F	0	40	40	60

the samples were tested as per standard ASTM D-790 [53]. For each sample, a total of five composites was tested to take average values. The flexural modulus of a composite sample was calculated by drawing a tangent line for the stress–strain curve; hence it is also called the tangent modulus. In general, the flexural modulus is defined as the capacity of a composite sample to deform when an external load acts on it. The flexural test samples were analyzed by using SEM.

Interlaminar Shear Test. To calculate the interlaminar shear strength of composite material, a universal testing machine with a full load capacity of 10 KN was used. To test the samples, the composite was fixed on both ends in between the jaws of the testing machine, and the load was applied in an axial direction from one end and the lower end was fixed. The samples were tested as per standard ASTM D-3846 [54]. The rectangular samples had a notch cut on both sides, 6.4 mm apart. The notch depth was considered half of the thickness of the sample on each side of the composite's specimen. The shear strength was measured only when failure occurred between the notches. The load acting on the composite gradually increased until the composite broke.

Wettability Test. The wettability properties of composite material were tested by using the contact angle measurement process. This provides quantitative information on interfacial energy between a fluid drop and composite surface [55]. A variety of fluids, such as spirit, water, and glycerin are used to measure the contact angle of a solid surface, of which, water is the most popular. Wettability indicates the examination of contact angles that regulate the wetting degree between a solid surface and a liquid drop. If the contact angle is less than 90° it means the solid surface has a hydrophilic nature and if it is more than 90° it has a hydrophobic nature. There are several factors that can affect the contact angle such as temperature, atmospheric conditions, type of liquid, the viscosity of the liquid, and time to take the measurement [56, 57]. In the present work, all the samples were tested in a dark room to avoid light, and dimensions were considered as per ASTM testing standards. The composite surfaces were cleaned properly before forming the water bubble. The contact angle values were measured at various positions on the composite surface.

SEM Energy Dispersive X-ray Spectrum Analysis. A scanning electron microscope (SEM) from Bruker, Billerica, USA was used to investigate the morphology of both the composite surface and the fractured sample. The element composition the surface of each composite sample and the facture position were evaluated by using SEM–EDX spectroscopy. The error % connected with the elemental composition analysis was evaluated.

Numerical Analysis. Numerical analysis of natural fiber composites was carried out by the using micromechanical method i.e., the Halpin–Tsai approach. Although there are many methods available to predict theoretical analysis, these are easy to apply to natural fiber composites. To calculate the elastic properties the basic assumptions must be considered as the composite sample was isotropic and homogeneous, as well as being linear elastic in nature [58, 59]. We modified the equations to analyze the hybrid fiber composite elastic properties. The modified equations to calculate the elastic properties are given below:

Halpin–Tsai method [60–62]

$$E = \frac{E_m\left(1 + \zeta\eta V_f\right)}{1 - \eta V_f} \tag{5.1}$$

where η is

$$\eta = \frac{E_f - E_m}{E_f + \zeta E_m} \tag{5.2}$$

By substituting the η value in Eq. (5.3) the final equation will be as follows:

$$E = \frac{E_m + \zeta V_f E_m\left(E_f - E_m\right)}{E_f + \zeta E_m - V_f\left(E_f - E_m\right)} \tag{5.3}$$

where E_f and E_m are the Young's modulus of fiber and matrix material, respectively. V_f and V_m are volume fractions of fiber and matrix material, respectively. ζ is an arbitrary variable, also called an empirical factor considered based on the assumptions. The ζ values are taken as 0, 1, 2, and so on. The modified Halpin's equation is as follows:

$$E = \frac{E_m + \zeta E_m(V_{f1} + (V_{f2})\left(E_{f1} + E_{f2} - E_m\right)}{E_{f1} + E_{f2} + \zeta E_m - (V_{f1} + V_{f2})\left(E_{f1} + E_{f2} - E_m\right)} \tag{5.4}$$

E_{f1}, E_{f2}, and E_m are Young's modulus of hemp fiber and flax fiber, and the matrix material, respectively. V_{f1}, V_{f2}, and V_m are the volume fractions of hemp fiber, flax fiber, and matrix material, respectively. In the present work, an attempt has been made to check the extent of the modified Halpin's equation to the natural fiber hybrid composites. The test was conducted by assigning properties such as Poisson's ratio, young's modulus, density, bulk modulus, and shear modulus to the material. The values are given in Table 5.2. The mechanical properties such as elastic modulus and displacement were determined and the results were compared with the theoretical and experimental results.

Table 5.2 Mechanical properties of the materials

Material	Density (g/cm^3)	Poisson's ratio	Young's modulus (GPa)	Shear modulus (GPa)	Bulk modulus (GPa)
Epoxy resin	1.10	0.29	3.35	2.10	1.408
Flax	1.4–1.5	0.41	60	21.27	1.11
Hemp	1.48	0.39	70	25.18	1.06
Ecopoxy	1.13	0.23	2.28	0.98	1.26

5.3 Results

Flexural properties. The flexural strength and flexural modulus of hemp and flax hybrid composites with various weight fractions in the presence of epoxy resin matrix and ecopoxy resin matrix are shown in Fig. 5.2. In both cases, hybrid composites showed superior flexural properties to those of individual fiber composites. This reflects the impact of hybridization on composite materials. It was observed that the epoxy-matrix-based composites exhibited greater flexural strength and flexural modulus than the ecopoxy matrix composites. This can be attributed to poor bonding and adhesion properties between the ecopoxy matrix and the reinforcement material. The epoxy matrix composites 25H/15F exhibited the highest flexural strength and modulus and the lowest was observed in the 0 h/40F composites, whereas in ecopoxy matrix composites, 25H/15F hybrid composites showed the highest values and 40H/0F composites showed the lowest values. These results proved that the matrix material plays a crucial role in the mechanical performance of the composite material.

Scanning electron microscope analysis of flexural test specimens of epoxy matrix composites is illustrated in Fig. 5.2. The presence of broken of fibers and fiber shrinkage in individual fiber composites can be observed (Fig. 5.3a, e). This lowers the strength of the composite, whereas in hybrid composites (Fig. 5.3b–d) strong bonding between reinforcement and matrix material was observed. This strong adhesion property of composites improves the stress distribution inside the composite material and improves the strength of a composite. In addition, minor cracks were observed in hybrid composites.

Interlaminar Shear Properties. The experimental results of epoxy and ecopoxy-based hybrid composites obtained from interlaminar shear tests are presented in Fig. 5.4. The shear strength of a composite is one of the crucial parameters in defining the capability of a composite to withstand delamination damage. From the results, it was clear that the interlaminar shear strength properties of hybrid composites were superior to those of the pure fiber (40H/0F and 0H/40F) composites in both cases. This indicates the strong bonding properties between the layers of the hybrid composite compared to the pure composite. The epoxy-based 25H/15F hybrid composite showed the highest shear strength at 13.01 MPa and the ecopoxy-based 40H/0F composite exhibited the lowest at 5.67 MPa.

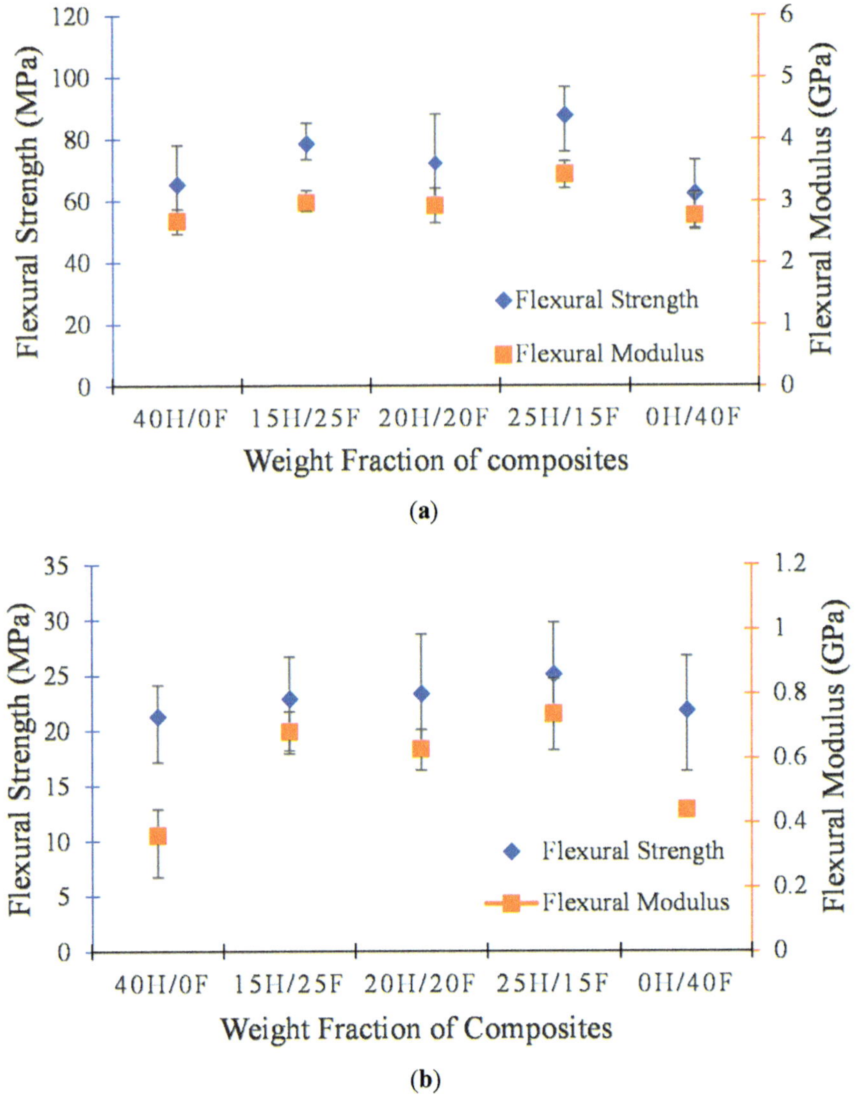

Fig. 5.2 Flexural strength and modulus of H/F hybrid composites with epoxy **a** and ecopoxy **b** matrices

Wettability Analysis. To analyze the wettability properties of a composite material a contact angle measurement was conducted on the hemp and flax fiber composites and the results are illustrated in Fig. 5.5. To understand the surface property of composite material, it is necessary to find the wetting nature of a composite [63]. It is easy to determine whether a composite is exhibiting a hydrophilic or a hydrophobic surface property by conducting the contact angle measurement analysis. It was found that

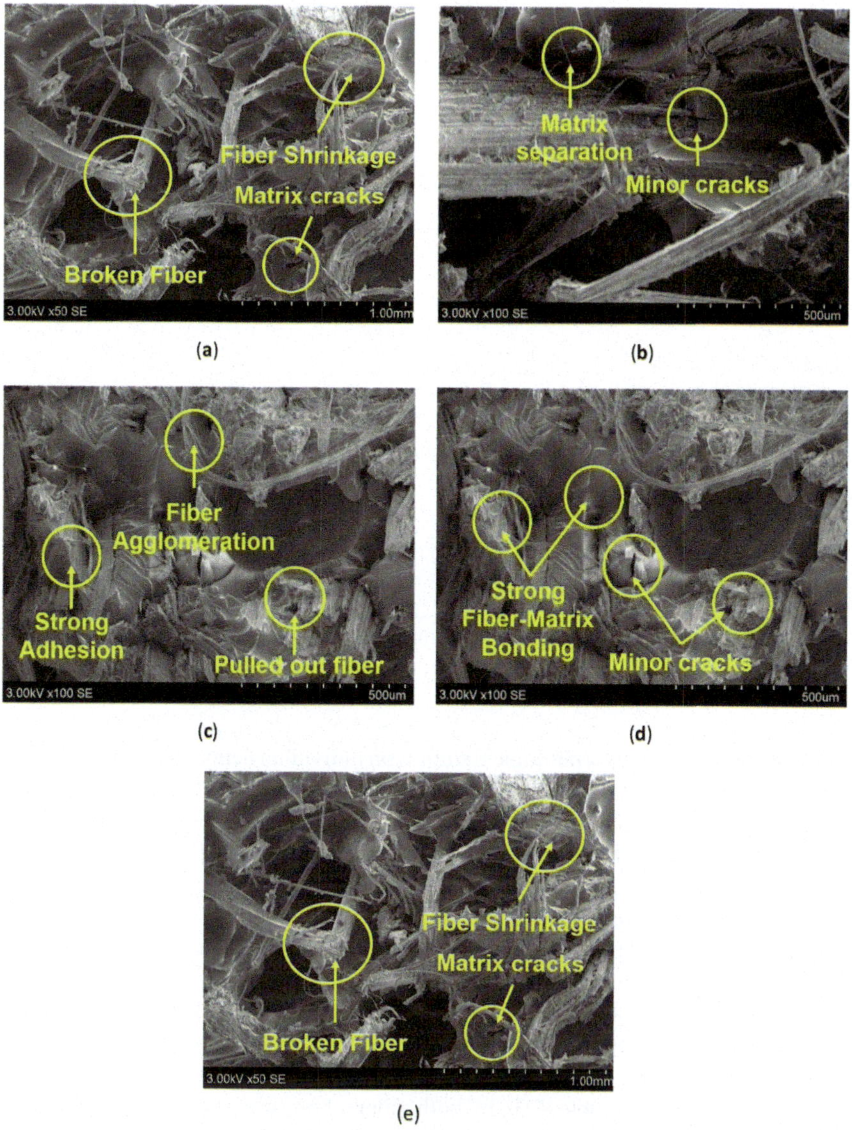

Fig. 5.3 SEM analysis of flexural tested hemp- and flax-fiber-reinforced epoxy matrix composites. **a** 40H/0F, **b** 15H/25F, **c** 20H/20F, **d** 25H/15F, and **e** 0H/40F

both epoxy and ecopoxy composites showed contact angles of <90° which meant that all composites exhibited hydrophilic surface nature. In both cases, hybrid composites showed a similar contact angle whereas pure composites exhibited a slight variation in it.

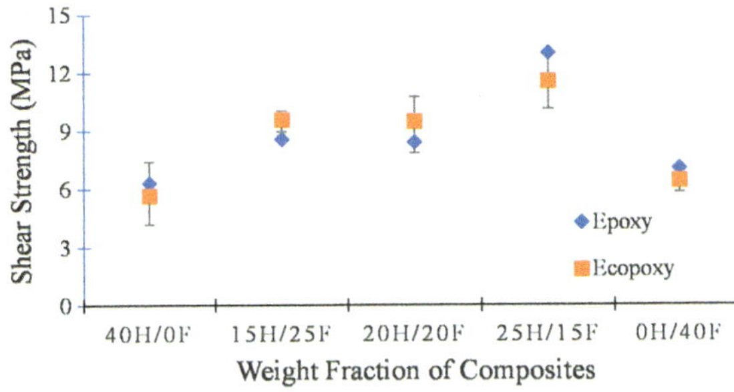

Fig. 5.4 Interlaminar shear strength of epoxy and ecopoxy composites with various weight fractions

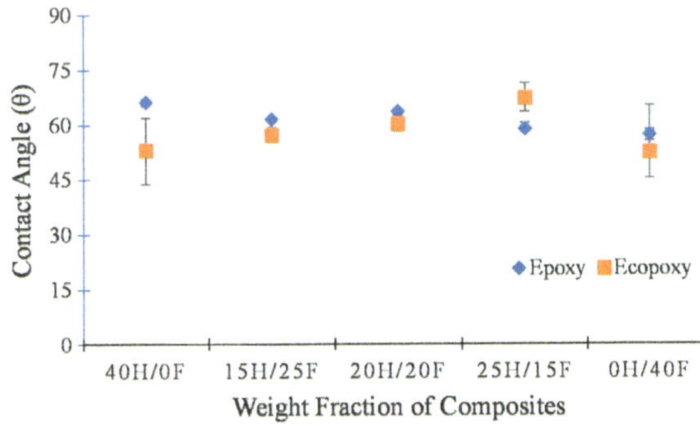

Fig. 5.5 Contact angle measurement of epoxy and ecopoxy composites with various weight fractions

SEM–EDX Analysis. Morphological studies on individual hemp- and individual flax-fiber-reinforced epoxy composites at the surface position and the fracture position were carried out by using the scanning electron microscope energy dispersive X-ray spectroscopy and the results are shown in Figs. 5.6 and 5.7. From SEM results it was observed that the presence of defects such as void content, fiber breakage, and matrix cracks were greater at the fracture position than at the surface position. The reasons for this could be poor reinforcement–matrix interactions and agglomeration of the fibers in the composite material.

Energy-dispersive X-ray (EDX) spectrum analysis is typically preferred to ascertain the chemical composition and elemental analysis of composite materials. The chemical composition and error percentage of hemp and flax fiber epoxy composites are presented in Table 5.3. In both fiber composites, carbon exhibited the highest percentage followed by oxygen, with the percentage of carbon being more pronounced at the surface than at the fracture location. Conversely, the atomic percentage of oxygen was higher at the fracture location compared to the surface location. At the surface location, only carbon, oxygen, chlorine, and a minor amount of sodium were detected, while at the fracture location, calcium, aluminum, and potassium were additionally observed alongside the chemical compositions present at the surface of the epoxy-based composite material.

Morphological analyses of individual hemp- and individual flax-fiber-reinforced ecopoxy matrix composites at both the surface and fracture positions were conducted

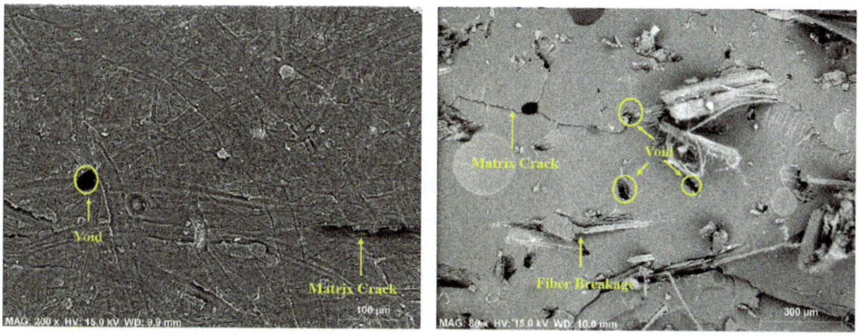

(a) (b)

Fig. 5.6 SEM EDX analysis on hemp-fiber-reinforced epoxy composite at surface **a** and fracture **b** positions

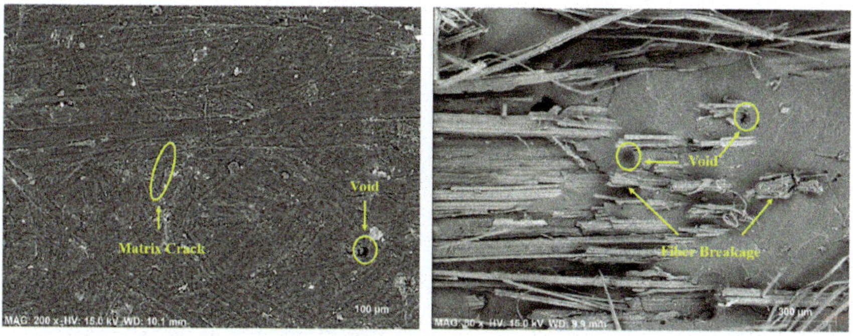

(a) (b)

Fig. 5.7 SEM EDX analysis of flax fiber reinforced epoxy composite at surface **a** and fracture **b** positions

Table 5.3 Chemical composition of hemp and flax fiber epoxy composites

Element	Hemp surface		Hemp fracture		Flax surface		Flax fracture	
	Atom. C (%)	Error (%)	Atom. C (%)	Error (%)	Atom. C (%)	Error (%)	Atom. C (%)	Error (%)
Carbon	73.74	21.6	69.97	20.3	74.66	21.9	59.98	21.8
Oxygen	26.15	10.4	29.76	11.4	25.09	10.0	39.57	14.7
Chlorine	0.11	0.0	0.10	0.0	0.15	0.0	0.18	0.0
Calcium	–	–	0.09	0.0	–	–	0.14	0.0
Aluminum	–	–	0.08	0.0	–	–	–	–
Sodium	–	–	–	–	0.10	0.0	–	–
Potassium	–	–	–	–	–	–	0.13	0.0

utilizing SEM–EDX spectroscopy, with the findings illustrated in Figs. 5.8 and 5.9. The SEM results indicated that, analogous to epoxy composites, defects such as void content, fiber breakage, and matrix cracks were more prevalent at the fracture position compared to the surface position. Notably, ecopoxy composites exhibited a higher incidence of defects than epoxy composites. These defects contribute to a reduction in composite strength, ultimately impacting the mechanical performance of the composite material.

The chemical composition and error % of the hemp and flax fiber ecopoxy composites are given in Table 5.4. In both fiber composites, oxygen had the highest percentage followed by carbon, The percentage of oxygen was higher at the fracture than at the surface position, whereas the atomic% of carbon was more at the surface position than the fracture position. Only carbon, oxygen, chlorine, and a small amount of silicon were observed at the surface position, but at the fracture position, carbon, oxygen, chlorine, silicon, and calcium were observed.

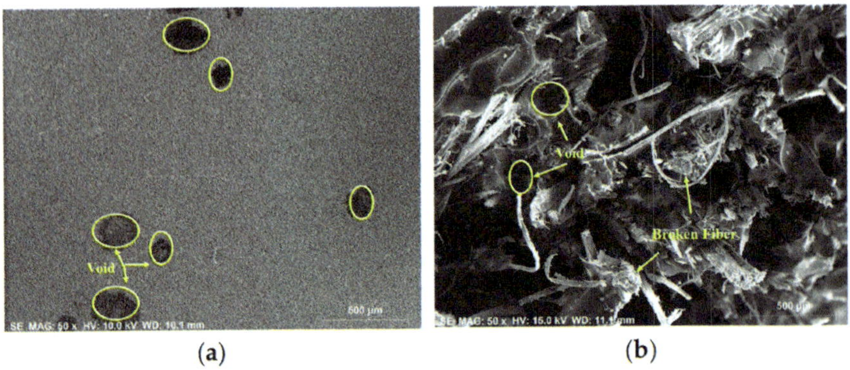

Fig. 5.8 SEM EDX analysis on hemp fiber reinforced ecopoxy composite at surface **a** and fracture **b** positions

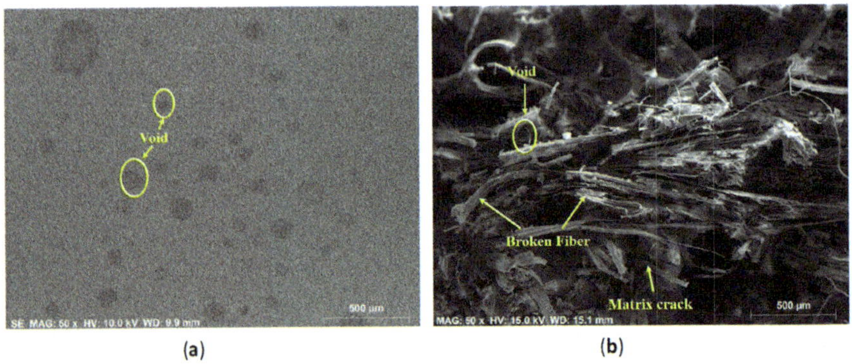

Fig. 5.9 SEM EDX analysis on flax fiber reinforced ecopoxy composite at surface **a** and fracture **b** positions

Table 5.4 Chemical composition of hemp and flax fiber ecopoxy composites

Element	Hemp surface		Hemp fracture		Flax surface		Flax fracture	
	Atom. C (%)	Error (%)	Atom. C (%)	Error (%)	Atom. C (%)	Error (%)	Atom. C (%)	Error (%)
Carbon	48.11	23.6	40.76	16.8	48.76	25.2	37.42	35.0
Oxygen	51.30	20.6	58.97	20.3	51.07	19.6	62.12	22.7
Chlorine	0.39	0.1	0.17	0.0	0.17	0.1	0.27	0.1
Silicon	0.20	0.1	–	–	–	–	–	–
Calcium	–	–	0.10	0.0	–	–	0.19	0.1

Overall, the defect percentage was greater in the ecopoxy composites than in the epoxy composites; hence, the strength of the composite was decreased in the ecopoxy case. Due to the biological nature of the ecopoxy resin material the percentage of oxygen was greater in ecopoxy composites than in the epoxy composites.

Numerical Analysis. Numerical analysis was carried out by using the modified Halpin's method. The values are presented in Figs. 5.10 and 5.11. In both cases, experimental values and numerical values were not close to each other and a large variation was observed when using Halpin's method. Halpin's method gave the higher data for all the hybrid composites investigated. This means that the modified Halpin's equations for calculation of elastic properties of hybrid-fiber-reinforced epoxy and ecopoxy composites gave higher bound results. Individual fiber composites showed the lowest elastic properties in both cases.

The data presented in Tables 5.5 and 5.6 shows the % error calculated for the epoxy and ecopoxy-matrix-based hybrid fiber composite's elastic properties. The error % was calculated by taking the ratio of the difference between theoretical values and experimental results upon the experimental results. It was observed that the error % calculated by using Halpin's method was far higher than the other methods. This indicates that the modified Halpin–Tsai theoretical calculations cannot be applied to natural fiber hybrid composites. It was observed that the elastic properties obtained from the numerical models were slightly higher than the experimental results in both epoxy and ecopoxy cases. This can be attributed to the fabrication errors and

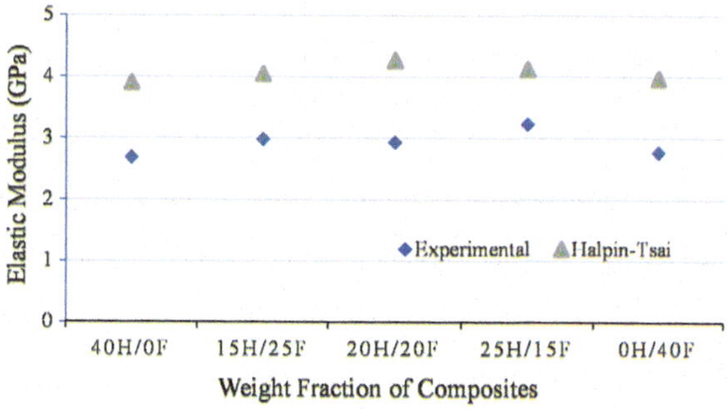

Fig. 5.10 Elastic properties of epoxy-based hybrid composites by various methods

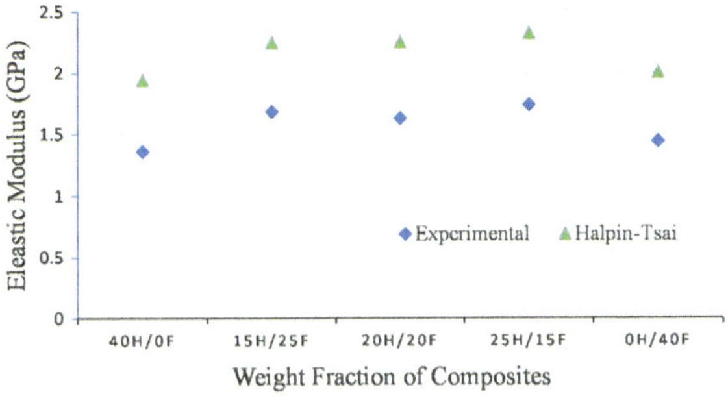

Fig. 5.11 Elastic properties of ecopoxy-based hybrid composites by various methods

environmental conditions during the experimentation process whereas in theoretical predication, the outside factors are not considered.

The percentage error was calculated for epoxy and ecopoxy hybrid composites. The error was calculated by taking the ratio of the difference between numerical values and experimental results. The maximum error % of elastic properties was observed in pure composites such as 0H/40F and 40H/0F in both epoxy and ecopoxy hybrid composites and the lowest was observed in 25H/15F epoxy hybrid composites. In the case of hybrid composites, the error % was similar whereas in pure composites it was different.

Table 5.5 Comparison of elastic properties of epoxy composites

Epoxy composites	Experimental	Halpin–Tsai method	Error (%)
40H/0F	2.683	3.896	45.21
15H/25F	2.981	4.042	35.59
20H/20F	2.934	4.264	45.33
25H/15F	3.226	4.118	27.65
0H/40F	2.772	3.968	43.15

Table 5.6 Comparison of elastic properties of ecopoxy composites

Ecopoxy composites	Experimental	Halpin–Tsai method	Error (%)
40H/0F	1.361	1.942	42.69
15H/25F	1.683	2.248	33.57
20H/20F	1.628	2.254	38.45
25H/15F	1.737	2.324	33.79
0H/40F	1.442	2.008	39.25

5.4 Discussion

The majority of prior research endeavors have been conducted on natural fiber-reinforced thermosetting or thermoplastic hybrid composites. Natural fibers, including banana, okra, jute, bamboo, and pineapple, were utilized as constituents of composite materials [64–66]. The current research primarily concentrated on examining the influence of matrix material on the mechanical properties of composite materials. To distinguish between matrix materials, epoxy and ecopoxy (bio-based resin) were employed as matrix materials. It was observed that hemp- and flax-fiber-reinforced epoxy composites exhibited superior properties compared to ecopoxy hybrid composites. Furthermore, hybrid composites demonstrated enhanced properties relative to individual fiber composites. The findings indicated that epoxy-based hybrid composites possessed flexural strength properties ranging from 58 to 80 MPa, whereas previous research documented a range between 45 and 64 MPa. Additionally, significant improvements were observed in the interlaminar shear properties. Based on contact angle measurement analysis, earlier results revealed that all natural fiber composites displayed hydrophilic surface properties (contact angle <70°). In contrast, the present research indicated that ecopoxy hybrid composites exhibited a contact angle approaching 90°, suggesting a tendency towards hydrophobic surface properties.

5.5 Conclusions

In the present chapter, the influence of matrix material on the mechanical properties of composite materials was examined by comparing epoxy-resin-matrix natural fiber composites with ecopoxy (bio-resin)-resin-matrix natural hybrid fiber composites. The findings are outlined below. Epoxy-matrix-based hybrid composites demonstrated superior mechanical properties compared to ecopoxy-matrix-based hybrid composites. It was noted that the mechanical properties of epoxy-resin hybrid composites were nearly twice as great as those of ecopoxy composites. In both instances, pure composites exhibited the weakest mechanical properties, such as flexural strength and interlaminar shear strength, while hybrid composites displayed the highest strength and shear properties. For epoxy composites, the maximum flexural and shear strengths were observed at 78 and 13 MPa, respectively, whereas for ecopoxy composites, these values were 30 and 10 MPa, respectively. The wettability properties of natural fiber hybrid composites were evaluated through contact angle measurements. The results indicated that all composites exhibited hydrophilic ($\theta <$ 90°) surface properties. The morphological characteristics and chemical composition within the composites were analyzed using SEM-RDX spectrum analysis.

To validate this research work, a numerical analysis was carried out including the experimental analysis. Theoretical studies were carried out to determine the elastic properties based on the modified Halpin–Tsai method. It was observed that the error

% was less with the numerical methods but in the case of Halpin's method, it was more. Hence Halpin's method is not suitable for predicting the elastic properties of natural fiber polymer composites.

Overall epoxy-based natural hybrid composites showed improved mechanical properties compared to ecopoxy-based hybrid composites. Hence the hemp and flax fiber reinforced epoxy matrix composites could be a potential replacement for synthetic fiber composites in various industries such as the automobile, construction, microfluidics, and biomedical industries.

References

1. Atmakuri A, Palevicius A, Vilkauskas A, Janusas G (2022) Numerical and experimental analysis of mechanical properties of natural-fiber-reinforced hybrid polymer composites and the effect on matrix material. Polym Basel MDPI. 14(13):1–18. https://doi.org/10.3390/polym14132612
2. Bhoopathi R, Deepa C, Sasikala G, Ramesh M (2015) Experimental investigation on mechanical properties of hemp-banana-glass fiber reinforced composites. Appl Mech Mater 766–767:167–172
3. Gokuldass R, Ramesh R (2019) Mechanical strength behavior of hybrid composites tailored by glass/kevlar fibre-reinforced in nano-silica and micro-rubber blended epoxy. SILICON 11:2731–2739
4. Ramakrishnan G, Ramnath BV, Elanchezhian C, Kumar AA, Gowtham S (2019) Investigation of mechanical behaviour of basalt-banana hybrid composites. SILICON 11:1939–1948
5. Sinha AK, Narang HK, Bhattacharya S (2017) Mechanical properties of natural fibre polymer composites. J Polym Eng 37:879–895
6. Väisänen T, Batello P, Lappalainen R, Tomppo L (2018) Modification of hemp fibers (*Cannabis sativa* L.) for composite applications. Ind Crops Prod 111:422–429
7. Pandey JK, Ahn SH, Lee CS, Mohanty AK, Misra M (2010) Recent advances in the application of natural fiber based composites. Macromol Mater Eng 295:975–989
8. Bodros E, Pillin I, Montrelay N, Baley C (2007) Could biopolymers reinforced by randomly scattered flax fibre be used in structural applications? Compos Sci Technol 67:462–470
9. Agung EH, Sapuan SM, Megat Ahmad MMH, Zaman DKM, Mustofa U (2011) Investigation of hardness of short abaca (Musa Textile Nee) fibre-reinforced high impact polystyrene (HIPS) composites by response surface methodology. Key Eng Mater 471–472:821–826
10. Mochane MJ, Mokhena TC, Mokhothu TH, Mtibe A, Sadiku ER, Ray SS, Ibrahim ID, Daramola OO (2019) Recent progress on natural fiber hybrid composites for advanced applications: a review. Exp Polym Lett 13:159–198
11. Li Y, Hu C, Yu Y (2008) Interfacial studies of sisal fiber reinforced high density polyethylene (HDPE) composites. Compos A Appl Sci Manuf 39:570–578
12. Sanjay MR, Madhu P, Jawaid M, Senthamaraikannan P, Senthil S, Pradeep S (2018) Characterization and properties of natural fiber polymer composites: a comprehensive review. J Clean Prod 172:566–581
13. Singha AS, Thakur VK (2008) Synthesis and characterization of pine needles reinforced RF matrix based biocomposites. E-J Chem 5:1055–1062
14. Messana A, Airale AG, Ferraris A, Sisca L, Carello M (2017) Correlation between thermo-mechanical properties and chemical composition of aged thermoplastic and thermosetting fiber reinforced plastic materials. Mater Werkst 48:447–455
15. Faruk O, Bledzki AK, Fink H, Sain M (2012) Biocomposites reinforced with natural fibers: 2000–2010. Prog Polym Sci 37:1552–1596

16. Azwa ZN, Yousif BF (2013) Characteristics of kenaf fibre/epoxy composites subjected to thermal degradation. Polym Degrad Stab 98:2752–2759
17. He H, Li K, Wang J, Sun G, Li Y, Wang J (2011) Study on thermal and mechanical properties of nano-calcium carbonate/epoxy composites. Mater Des 32:4521–4527
18. George J, Sreekala MS, Thomas S (2001) A review on interface modification and characterization of natural fiber reinforced plastic composites. Polym Eng Sci 41:1471–1485
19. Manteghi S, Mahboob Z, Fawaz Z, Bougherara H (2017) Investigation of the mechanical properties and failure modes of hybrid natural fiber composites for potential bone fracture fixation plates. J Mech Behav Biomed Mater 65:306–316
20. Ramadan N, Taha I, Hammouda R, Abdellatif MH (2017) Behaviour of hybrid SiC/jute epoxy composites manufactured by vacuum assisted resin infusion. Polym Polym Compos 25:333–344
21. Yusoff RB, Takagi H, Nakagaito AN (2016) Tensile and flexural properties of polylactic acid-based hybrid green composites reinforced by kenaf, bamboo and coir fibers. Ind Crops Prod 94:562–573
22. Nunna S, Chandra PR, Shrivastava S, Jalan A (2012) A review on mechanical behavior of natural fiber based hybrid composites. J Reinf Plast Compos 31:759–769
23. Dong C (2019) Mechanical properties of natural fibre-reinforced hybrid composites. J Reinf Plast Compos 38:910–922
24. Tang J, Swolfs Y, Longana ML, Yu H, Wisnom MR, Lomov SV, Gorbatikh L (2019) Hybrid composites of aligned discontinuous carbon fibers and self-reinforced polypropylene under tensile loading. Compos A Appl Sci Manuf 123:97–107
25. Kureemun U, Ravandi M, Tran LQN, Teo WS, Tay TE, Lee HP (2018) Effects of hybridization and hybrid fibre dispersion on the mechanical properties of woven flax-carbon epoxy at low carbon fibre volume fractions. Compos B Eng 134:28–38
26. Premkumar T, Siva I, Amico SC (2020) Inter and intralayer basalt hybrid effects on the static and vibrational behaviors of Brazilian curauá/basalt hybrid composite. Mater Today Proc 33:1212–1215
27. Dong C (2018) Review of natural fibre-reinforced hybrid composites. J Reinf Plast Compos 37:331–348
28. Marques G, Rencoret J, Gutiérrez A, del Río JC (2010) Evaluation of the chemical composition of different non-woody plant fibers used for pulp and paper manufacturing. Open Agric J 4:93–101
29. Komuraiah A, Kumar NS, Prasad BD (2014) Chemical composition of natural fibers and its influence on their mechanical properties. Mech Compos Mater 50:359–376
30. Wang X, Xie J, Zhang H, Zhang W, An S, Chen S, Luo C (2019) Determining the lignin distribution in plant fiber cell walls based on chemical and biological methods. Cellulose 26:4241–4252
31. Fiore V, Scalici T, Valenza A (2014) Characterization of a new natural fiber from *Arundo donax* L. as potential reinforcement of polymer composites. Carbohydr Polym 106:77–83
32. Chokshi S, Parmar V, Gohil P, Chaudhary V (2020) Chemical composition and mechanical properties of natural fibers. J Nat Fibers 17:1–12
33. Riccio A, Raimondo A, Saputo S, Sellitto A, Battaglia M, Petrone G (2018) A numerical study on the impact behaviour of natural fibres made honeycomb cores. Compos Struct 202:909–916
34. Karimah A, Ridho MR, Munawar SS, Adi DS, Damayanti RI, Subiyanto B, Fatriasari W, Fudholi A (2021) A review on natural fibers for development of eco-friendly bio-composite: characteristics, and utilizations. J Mater Res Technol 13:2442–2458
35. Pan Y, Zhong Z (2015) A micromechanical model for the mechanical degradation of natural fiber reinforced composites induced by moisture absorption. Mech Mater 85:7–15
36. Neves ACC, Rohen LA, Mantovani DP, Carvalho JPRG, Vieira CMF, Lopes FPD, Simonassi NT, da Luz FS, Monteiro SN (2020) Comparative mechanical properties between biocomposites of Epoxy and polyester matrices reinforced by hemp fiber. J Mater Res Technol 9:1296–1304

37. Sapuan SM, Leenie A, Harimi M, Beng YK (2006) Mechanical properties of woven banana fibre reinforced epoxy composites. Mater Des 27:689–693
38. Sharavanan R et al (2018) A review on natural fiber hybrid composites. Int J Mech Prod Eng Res Dev 8:943–948
39. Abedom F, Sakthivel S, Asfaw D, Melese B, Solomon E, Kumar SS (2021) Development of natural fiber hybrid composites using sugarcane bagasse and bamboo charcoal for automotive thermal insulation materials. Adv Mater Sci Eng 2021:2508840
40. Summerscales J, Virk A, Hall W (2013) A review of bast fibres and their composites: part 3—modelling. Compos A Appl Sci Manuf 44:132–139
41. Alhazmi WH, Jazaa Y, Mousa S, Abd-Elhady AA, Sallam HEM (2021) Tribological and mechanical properties of epoxy reinforced by hybrid nanoparticles. Lat Am J Solids Struct 18:e361
42. Singh S, Deepak D, Aggarwal L, Gupta VK (2014) Tensile and flexural behavior of hemp fiber reinforced virgin-recycled HDPE matrix composites. Proced Mater Sci 6:1696–1702
43. Moudood A, Rahman A, Huq NML, Öchsner A, Islam MM, Francucci G (2020) Mechanical properties of flax fiber-reinforced composites at different relative humidities: experimental, geometric, and displacement potential function approaches. Polym Compos 41:4963–4973
44. Zhong Y, Kureemun U, Tran LQN, Lee HP (2017) Natural plant fiber composites-constituent properties and challenges in numerical modeling and simulations. Int J Appl Mech 9:1750045
45. Behzad T, Sain M (2007) Finite element modeling of polymer curing in natural fiber reinforced composites. Compos Sci Technol 67:1666–1673
46. José da Silva L, Hallak Panzera T, Luis Christoforo A, Miguel Pereira Dur L, Antonio Rocco Lahr F (2012) Numerical and experimental analyses of biocomposites reinforced with natural fibres. Int J Mater Eng 2:43–49
47. Sathishkumar GK, Mohamed Akheel M, Ibrahim M, Rajkumar G, Karpagam R, Gopinath B (2021) Experimental and finite element analysis of lignite fly ash on the mechanical properties of sisal-added polymer matrix composite using ANSYS workbench. J Nat Fibers 18:1–25
48. Prasanthi PP, Babu KS, Niranjan-Kumar MSR, Kumar AE (2021) Analysis of sisal fiber waviness effect on the elastic properties of natural composites using analytical and experimental methods. J Nat Fibers 18:1675–1688
49. Balasubramanian K, Rajeswari N, Vaidheeswaran K (2019) Analysis of mechanical properties of natural fibre composites by experimental with FEA. Mater Today Proc 28:1149–1153
50. Gupta MK, Ramesh M, Thomas S (2021) Effect of hybridization on properties of natural and synthetic fiber-reinforced polymer composites (2001–2020): a review. Polym Compos 42:4981–5010
51. Khan MZR, Srivastava SK, Gupta M (2018) Tensile and flexural properties of natural fiber reinforced polymer composites: a review. J Reinf Plast Compos 37:1435–1455
52. Aji I, Zainudin E, Abdan K, Sapuan S, Khairul M (2013) Mechanical properties and water absorption behavior of hybridized kenaf/pineapple leaf fibre-reinforced high-density polyethylene composite. J Compos Mater 47:979–990
53. ASTM (2003) Standard test method for 3-point flexure tests on rigid and semi-rigid plastics as well as on fiber composites with long-fiber reinforcements. In: Annual book of ASTM standards; ASTM D-790. ASTM International, West Conshohocken, PA
54. ASTM D-3846-02 (2005) Standard test method for in plane shear strength of reinforced plastics. In: Annual book of ASTM standards. ASTM International, West Conshohocken, PA
55. Prakash S, Yeom J (2014) Interfaces in microfluidic and nanofluidic systems. Nano fluidics microfluid. Elsevier, Amsterdam, pp 39–86
56. Esmonde-White KA, Esmonde-White FWL, Morris MD, Roessler BJ (2014) Characterization of biofluids prepared by sessile drop formation. Analyst 139:2734–2741
57. Skedros JG, Knight AN, Clark GC, Crowder CM, Dominguez VM, Qiu S, Mulhern DM, Donahue SW, Busse B, Hulsey BI et al (2013) Scaling of Haversian canal surface area to secondary osteon bone volume in ribs and limb bones. Am J Phys Anthropol 151:230–244
58. Tham MW, Fazita MN, Abdul-Khalil HPS, Mahmud-Zuhudi NZ, Jaafar M, Rizal S, Haafiz MM (2019) Tensile properties prediction of natural fibre composites using rule of mixtures: a review. J Reinf Plast Compos 38:211–248

59. Zhao S, Zhao Z, Yang Z, Ke L, Kitipornchai S, Yang J (2020) Functionally graded graphene reinforced composite structures: a review. Eng Struct 210:110339
60. Facca AG, Kortschot MT, Yan N (2006) Predicting the elastic modulus of natural fibre reinforced thermoplastics. Compos A Appl Sci Manuf 37:1660–1671
61. Sathishkumar TP, Navaneethakrishnan P, Shankar S, Rajasekar R (2014) Mechanical properties and water absorption of short snake grass fiber reinforced isophthallic polyester composites. Fibers Polym 15:1927–1934
62. Nurihan O, Hisham MRM, Rasid ZA, Hassan MZ, Nor AFM (2019) The elastic properties of unidirectional bamboo fibre reinforced epoxy composites. Int J Recent Technol Eng 8:7187–7193
63. Marmur A (2008) From hygrophilic to superhygrophobic: theoretical conditions for making high-contact-angle surfaces from low-contact-angle materials. Langmuir 24:7573–7579
64. Bazan P, Mierzwiński D, Bogucki R, Kuciel S (2020) Bio-based polyethylene composites with natural fiber: mechanical, thermal, and ageing properties. Materials 13:2595
65. Pereira AL, Banea MD, Neto JSS, Cavalcanti DKK (2020) Mechanical and thermal characterization of natural intralaminar hybrid composites based on sisal. Polymers 12:866
66. Vijay C, Pramendra Kumar B, Sachin M (2018) An investigation on wear and dynamic mechanical behavior of jute/hemp/flax reinforced composites and its hybrids for tribological applications. Fibers Polym 19(2):403–415

Chapter 6
Investigation of Hemp and Flax Fiber-Reinforced EcoPoxy Matrix Biocomposites

Abstract Contemporary industries are predominantly concentrated on the advancement of bio-inspired hybrid natural fiber composites for use in lightweight biosensor chips, automotive, and microfluidic applications. In the present chapter, an investigation was conducted on the mechanical properties and morphological characteristics of alkaline (NaOH) treated hemp, flax, noil hemp, and noil flax fiber-reinforced ecopoxy bio-composites (Atmakuri et al. in Polym Basel MDPI 14:1–21, 2022 [1]). The specimens were fabricated utilizing the hand layup technique followed by the compression molding process. Two distinct sets of composites were produced with varied weight fractions. The specimens underwent testing to assess their mechanical properties, including flexural strength, interlaminar shear strength, and contact angle measurement. The treated fibers were examined through optical microscopy and Fourier Transform Infrared spectroscopy (FTIR). Morphological analyses, such as porosity and fracture mechanisms, were conducted utilizing scanning electron microscopy and SEM–EDX spectroscopy. The findings indicated that the incorporation of ecopoxy resin augments the mechanical properties, and the NaOH treatment enhances fiber properties by removing impurities. Overall, hemp/flax fiber bio-composites exhibited superior mechanical properties compared to noil hemp/noil flax fiber bio-composites. The effects of hybridization were distinctly discernible in both instances. It was noted that hybrid composites demonstrated enhanced properties compared to single fiber (individual fiber) composites.

6.1 Introduction

In contemporary times, numerous initiatives have been undertaken to develop innovative and environmentally sustainable materials for a variety of applications, driven by heightened ecological and environmental awareness [2]. Natural fibers are being explored as reinforcement in polymer composites due to the high cost of synthetic fibers and their detrimental environmental impact. Composites based on natural fibers provide environmental advantages by diminishing reliance on petroleum-based resources in current technologies [2, 3]. Biocomposites consist of natural fibers such

A. Atmakuri et al., *Fiber-Based Hybrid Composites for Engineering*, Springer Series on Polymer and Composite Materials, https://doi.org/10.1007/978-3-031-93144-4_6

as hemp, flax, sisal, Carlota, banana, pineapple, palm, kenaf, bamboo, jute, coir, and sugarcane, combined with either biodegradable or non-biodegradable polymers [4, 5]. The natural fiber-reinforced composites have garnered significant interest from researchers due to their abundance, cost-effectiveness, environmentally friendly nature, and applicability in lightweight applications. Natural fiber composites find applications in the automotive, marine, sports equipment, construction, aerospace, and microfluidic sectors [6–8]. Hemp and flax fibers, as renewable resources, exhibit promising mechanical properties for their use as reinforcement in composites across a broad scope of applications. Numerous research studies have explored the impact of fiber content and orientation on the mechanical properties of various natural fiber-reinforced polymer composites, encompassing both thermoplastic and thermosetting types [9–13]. Typically, an increase in fiber content enhances the flexural, torsional, and tensile strengths of these composites to a certain extent, beyond which these properties decline rapidly. Nevertheless, due to weak interfacial adhesion between reinforcement and matrix material, the mechanical properties of these composites may fall significantly short of their optimal values. Consequently, researchers are exploring methods to enhance the bonding between reinforcement and matrix materials through the application of various chemical treatment techniques [14, 15]. Such chemical treatments can modify the interface by interacting with both fiber and resin materials on the composite surface. Among the chemical treatment processes studied, alkaline treatment is the most prevalent coupling agent used to enhance bonding between fiber and matrix materials [16].

The hybridization of composites plays a significant role in advancing material properties, as demonstrated in recent years. Numerous studies have concentrated on enhancing material properties by developing new combinations, such as integrating two distinct natural fibers, combining natural fibers with synthetic fibers, and pairing two different synthetic fibers within the same matrix material. These composites have been examined to ensure improvements in mechanical properties over single fiber composites [17, 18]. Several types of hybrid composites exist, characterized by variations in size, filler material, and fiber orientation. The process of hybridization can mitigate the limitations of a single component by incorporating an alternative type of fiber material. Generally, the hybrid effect varies with the mechanical and chemical properties of the fibers, typically resulting in a lesser effect compared to individual fiber composites [19, 20]. Hybrid reinforcement with appropriate fiber selection can potentially deliver remarkable properties, meeting current demands for thermosetting matrix composites [21, 22]. Over recent decades, extensive research has been conducted on the hybridization of synthetic fibers and plant fibers (leaf, bast, seed). Nonetheless, natural fiber reinforced hybrid composites are less frequently investigated, with recent studies primarily focusing on the combination of natural fibers with artificial fibers. Exemplary composites may incorporate carbon, glass, and aramid in conjunction with hemp, flax, banana, and pineapple [23, 24]. However, the global increase in the utilization of synthetic polymer materials has underscored the importance of addressing environmental issues such as waste treatment, waste transfer management, and the contamination caused by burning. These concerns highlight the deleterious effects on environmental and climate change issues [25].

Numerous researchers have identified that the incorporation of natural fibers with a thermosetting polymer matrix results in superior properties compared to single fiber composites [26, 27]. Ramesh et al. [28] examined the mechanical and thermal properties of epoxy composites reinforced with banana-pineapple fibers. The findings indicated that the hybrid composites consisting of banana and pineapple fibers demonstrated enhanced mechanical properties, suggesting their potential application in light-load conditions pertinent to automotive and construction sectors. Although the flame retardancy properties assessed through thermogravimetric analysis were below expectations, these could be improved via appropriate chemical treatment processes. Zalinawati et al. [29] explored the impact of fiber treatment on the mechanical properties of burl palm fiber epoxy composites. Their study involved treating the burl palm fiber with a sodium hydroxide solution, followed by a fabrication process using the hand layup technique. The results indicated that the treated fiber composites exhibited superior mechanical properties compared to their untreated counterparts. It was observed that the untreated fiber composites had greater thickness swelling, attributed to higher water absorption when exposed to moist conditions. Torres-Arellano et al. [30] focused on the mechanical properties of natural fiber-reinforced biobased epoxy resins. Their investigation encompassed the fabrication of composites via a vacuum-assisted resin infusion methodology. The findings demonstrated that the integration of biobased epoxy resins within the composite matrix resulted in enhanced mechanical properties, with the jute fiber showing superior performance among the fibers analyzed. Jamshaid et al. [31] conducted an examination of the mechanical and electrical properties of natural fiber-reinforced bisphenol F epoxy-based bio-composites, utilizing jute, sisal, coconut, and sugarcane fibers. The outcomes indicated that among all the natural fiber composites, sisal/jute hybrid composites manifested superior mechanical properties relative to the coconut/sugarcane composites. The results derived from the fabricated specimens suggest that natural fiber biocomposites could pave the way for novel applications in electrical goods and as an environmentally sustainable alternative. Rajak et al. [32] explored the flexural mechanical properties of natural fiber reinforced polymer composites in conjunction with their applications. Within this synopsis, the authors endeavored to elucidate the benefits and diverse applications of natural fiber-reinforced composites. Terwadkar et al. [33] scrutinized the mechanical properties of epoxy hybrid composites reinforced with banana and kenaf fibers. In this section, the influence of hybridization was assessed by contrasting treated and untreated fibers. The application of alkaline treatment was found to substantially affect the mechanical properties, with hybrid composites exhibiting superior characteristics. Neves et al. [34] conducted a comparative analysis of biocomposites composed of epoxy and polyester. A study on hemp fiber composites was undertaken, wherein fiber weight percentages were varied at 10%, 20%, and 30%. The results indicated that an increase in fiber content correlated with enhanced mechanical properties to a certain degree. Epoxy-based biocomposites demonstrated superior mechanical properties compared to polyester composites. Thiagamani et al. [35] examined the mechanical properties of biocomposites based on hemp and sisal fibers. The experimental findings suggest that variations in the stacking sequence of hybrid composites resulted in minor changes in mechanical properties. Notably, the hybrid composite with an intercalated arrangement exhibited the highest tensile modulus relative to its counterparts. Typical failure characteristics observed in the short beam test included inter-laminar shear cracks in the transverse direction, micro-buckling, and fiber rupture, underscoring a deficiency in chemical treatment of the reinforcements. Bazan et al. [36] explored biobased polyethylene composites integrated with flax, basalt, and coconut fibers in conjunction with wood flour particles. Findings revealed a positive impact on bio-polyethylene composites, with basalt fiber composites demonstrating the greatest strengthening properties. A review of these studies indicates that alkaline-treated natural fibers presented enhanced mechanical properties by improving the adhesion between the reinforcement and the matrix material. Despite the user-friendly nature of epoxy resin materials in composite fabrication, certain drawbacks persist, including non-biodegradability, susceptibility to chemical degradation, and lack of skin compatibility. To address these issues, the current research endeavors to develop bio-inspired resin materials.

This study introduces novel composites formulated from ecopoxy (bio epoxy) in conjunction with natural fibers including hemp, flax, noil hemp, and noil flax. Two distinct sets of hybrid composites (Hemp/Flax and Noil hemp/Noil flax) with varying weight fractions were developed, grounded in the principle of hybridization. Fabrication of these composites was executed utilizing the hand layup technique, succeeded by the compression molding technique. These composites underwent testing to ascertain various material properties such as contact angles, moisture absorption, and mechanical attributes. The failure mechanisms were examined through the employment of a scanning electron microscope. Surface properties were assessed via optical microscopy. Mechanical properties were scrutinized and subsequently compared. Furthermore, the research aimed to evaluate whether noil fibers (considered as primary waste post-processing) could serve as replacements for conventional fibers.

6.2 Materials and Methods

Hemp, flax, noil hemp, and noil flax fibers are utilized as reinforcement in the production of composite materials. These fibers, specifically hemp and flax, were manually extracted from the plants. Noil fibers represent processed fibers, which are the primary waste by-products of standard fibers. An NaOH solution, procured from Sigma Aldrich, is employed for chemical treatment purposes. The matrix material employed is ecopoxy, a bio-inspired resin with a bioactivity level of 36%, in conjunction with a hardener that functions as a cross-linking agent; both components are sourced from Ecopoxy, Poland. The mechanical properties and chemical characteristics of the individual fibers are presented in Table 6.1.

Extraction and chemical treatment of natural fibers. A critical component of the study involves the selection and extraction of fibers, which is subsequently followed by chemical treatment. Hemp fibers were extracted from the Cannabis sativa plant. Flax fibers were obtained from the bast located beneath the outer layer of the flax

Table 6.1 Hemp and flax fibers chemical composition and mechanical properties [37–40]

Properties	Hemp	Flax
Cellulose (%)	70–74	64–70
Hemicellulose (%)	21–24	16–18
Lignin (%)	3.7–5.7	2–2.2
Pectin (%)	7.3	2.3
Elongation (%)	1.6–4.0	1.2–1.6
Moisture (%)	6.2–12	8–12
Density (g/cm^3)	1.48	1.4
Young's modulus (GPa)	70	60–80
Microfibrillar angle (°)	2–6	5–8

plant stem. Noil fibers were derived from the primary waste generated during the processing of raw fibers. Following extraction, all fibers were cleansed using distilled water and subsequently dried to eliminate moisture. The fibers were then subjected to an alkaline treatment process to enhance the adhesion between the matrix and the reinforcement. During the chemical treatment, a 5% solution of NaOH was prepared and separately applied to the fibers, which were then subjected to a heat treatment process at a temperature of 80 °C for a duration of 4 h.

Materials Weight Percentage. The weight fraction of both fiber and resin material was considered as per the hybridization concept. In the current research work, a total of two sets of hybrid biocomposites were fabricated based on hemp (H), flax (F), noil hemp (NH), and noil flax (NF) natural fibers. The hybridization states that the composite material contains more than one reinforcement under the same material with a 0.4 wf.%. The theoretical calculations of weight and volume fractions of both reinforcement and matrix material are given in the following equations.

The Weight fraction of the fiber (Reinforcement) material is well-defined as,

$$W_f = \frac{\text{Weight of fibers}}{\text{Total weight}} \tag{6.1}$$

In consequence, the weight fraction of resin material (Matrix) material is defined as,

$$W_m = \frac{\text{Weight of matrix}}{\text{Total weight}} \tag{6.2}$$

Similarly,

$$W_m + W_f = 1 \tag{6.3}$$

The volume fraction of fiber (Reinforcement) material is given as,

$$V_f = \frac{\text{Volume of fibers}}{\text{Total Volume}} \tag{6.4}$$

In consequence, the volume fraction of resin material (Matrix) material is defined as,

$$V_m = \frac{\text{Volume of resin}}{\text{Total Volume}} \tag{6.5}$$

Similar to the previous equation,

$$V_m + V_f = 1 \tag{6.6}$$

Not that one can change from weight fraction over completely to the volume fraction as well as other way around. If the densities of fiber (ρ_f) and matrix (ρ_m) materials are definite weight of fiber and matrix, individually. Then the equation can be written as follows.

$$V_f = \frac{\frac{W_f}{\rho_f}}{\frac{W_f}{\rho_f} + \frac{W_m}{\rho_m}} \tag{6.7}$$

From Eq. (6.7), the weight fraction of fiber is written as,

$$W_f = \frac{V_f \rho_f}{V_f \rho_f + V_m \rho_m} \tag{6.8}$$

The weight proportions are given in the following Table 6.2.

Table 6.2 Composite weight fractions

Composites	Hemp fiber (%)	Flax fiber (%)	Total fiber Vol. (%)	Total resin Vol. (%)
40H/0F	40	0	40	60
25H/15F	25	15	40	60
20H/20F	20	20	40	60
15H/25F	15	25	40	60
0H/40F	0	40	40	60
	Noil hemp fiber (%)	Noil flax fiber (%)		
40NH/0NF	40	0	40	60
25NH/15NF	25	15	40	60
20NH/20NF	20	20	40	60
15NH/25NF	15	25	40	60
0NH/40NF	0	40	40	60

Fabrication of composites. Hemp/flax and Noil hemp/noil flax natural fiber reinforced hybrid composites were synthesized utilizing the hand layup method, subsequently followed by a compression molding process. The fiber materials were quantified in accordance with the weight fractions specified in the preceding Table 6.2 and subsequently positioned within a mold material. The molds are lined with non-stick paper to prevent direct contact between the resin and the mold material. Subsequent to the arrangement of the fibers within the mold, EcoPoxy resin solution was poured, and a roller was employed on the fibers to achieve uniform distribution; the materials were then compressed under a constant weight and allowed to undergo the curing process. The entire setup was left for 72 h to complete the curing process. Post-curation of the biocomposite panels was conducted following fabrication. The entirety of the composite materials fabrication process is depicted in Fig. 6.1. For the post-curation process, composite panels were placed within an oven at 70 °C for a duration of 6 h. Following thermal treatment, the samples were extracted from the panels in accordance with ASTM standards.

Interlaminar shear properties The interlaminar shear strength of composite material is the failure shear that occurs when a transverse force is applied to it. To measure the shear strength, all the samples were allowed for a notch cut as 6.4 mm apart from the center and half of the thickness of the composites sample before testing. Tinus Olsen H10KT (Horsham, PA, USA) was used for testing samples. The dimensions of the tested samples are considered as per ASTM D-3846 standards. The interlaminar shear strength of the fabricated composites is calculated by using the following equation. Where p is maximum load and A is failure shear area.

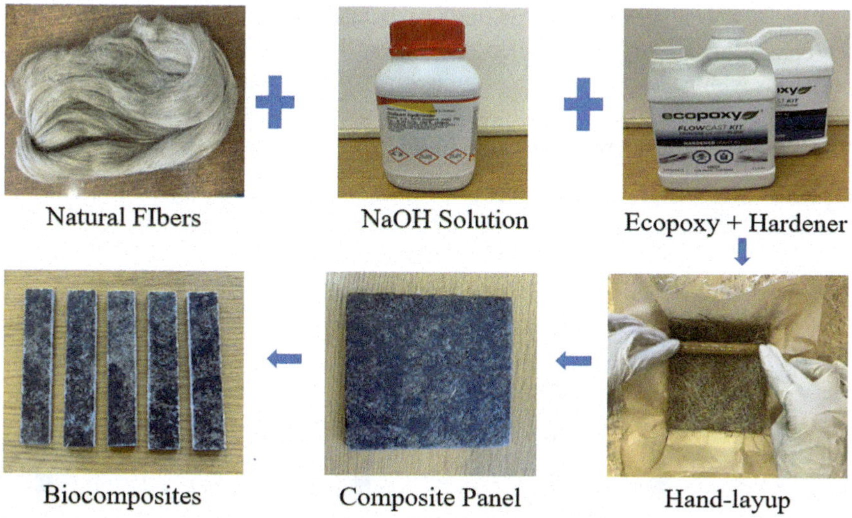

Fig. 6.1 Fabrication process of biocomposites

$$\sigma_S = \frac{P}{A}\text{MPa} \tag{6.9}$$

Flexural strength properties. A three-point bending test was performed on Tinus Olsen H10K to find out the flexural strength of the composites. All samples were tested as per ASTM D-790 standards. The following relation is used to calculate the flexural strength values where P is the maximum load, L is the span length, b is width, and d is the thickness of the composite sample

$$\sigma_f = \frac{3PL}{2bd^2}\text{ Mpa} \tag{6.10}$$

The flexural modulus is computed by dividing the change in stress by the corresponding change in strain and is represented by the slope of the initial straight-line segment of the stress–strain curve. As a result, the flexural modulus is defined as the ratio of stress to strain. The flexural modulus of composites is calculated by using the following relation where L is the length, m is the slope of the stress–strain curve, b is width, and d is the thickness of the composite.

$$E_B = \frac{L^3 m}{4bd^3}\text{ MPa} \tag{6.11}$$

Hydrophilic properties. The wettability of the composite surface is measured by using the contact angle measurement analysis. It is also possible to investigate the wetting of and dynamic contact angles on composite samples that are used as the test body. The main objective of this test is to determine the hydrophobicity of the composite material. A detailed description of the contact angle measurement apparatus with necessary parts is given in Fig. 6.2. The schematic representation of the contact angle on the composite surface is shown in Fig. 6.3 and the test specimen standards were considered as per ASTM D-7334 standards.

Optical microscopic analysis. The modification of fiber surfaces is frequently accomplished through physical or chemical techniques to enhance adhesion between

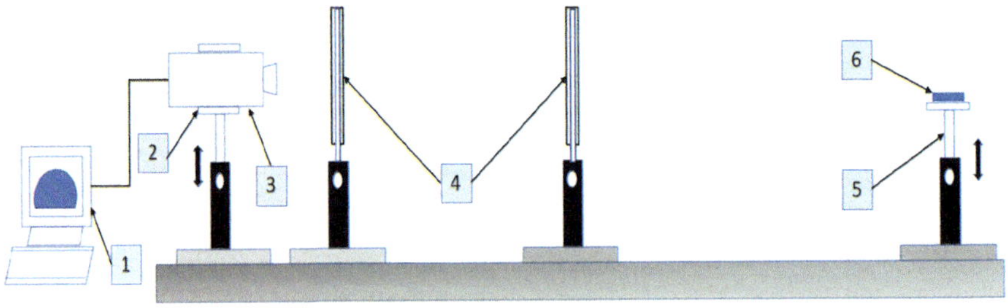

Fig. 6.2 Schematic representation of contact angle test apparatus containing (1) computer, (2) movable holder (3) camera, (4) concave lenses, (5) composite rest, (6) composite

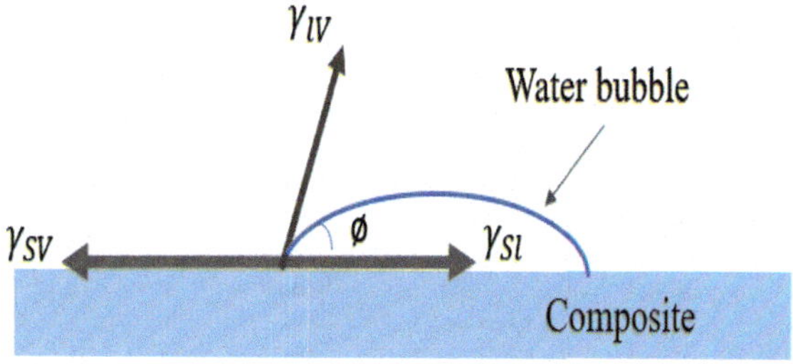

Fig. 6.3 Contact angle formed on composite surface with interfacial tensions of the three phases

the reinforcement and matrix. This leads to a decrease in moisture absorption and alterations in the surface characteristics of the fibers. Alterations in the surface morphology of fibers are crucial for evaluating their interaction with polymer matrices within composite materials. The impact of chemical treatments on natural fibers is examined utilizing an optical microscope.

FTIR analysis. The surface functional groups analysis was performed using Fourier Transform Infrared spectrometer (FTIR, Bruker tensor 27, Germany). For the FTIR analysis samples firstly were dried at 105 °C for 24 h according to ISO 579 standard. After the preparation, the sample was placed on a horizontal ZnSe crystal plate with a 45° incidence angle and 10 internal reflections and pressed with a fixator to avoid any gap between the plate and sample. A background reference spectra were obtained before every experiment, to eliminate the influence of the atmosphere. The scanning time of every wavelength is 32 times, while the wavelength interval is from 650 to 4000 cm^{-1}. The resolution of the detector is 4 cm^{-1}.

SEM and EDX analysis. The morphological studies, biocomposites failure mechanism, and the chemical composition of the biocomposites were investigated by using the SEM (scanning electron microscope) from Hitachi along with an EDX (energy dispersive X-Ray) spectrometer from Bruker, Billerica, USA.

6.3 Results and Discussion

Interlaminar shear strength test. The interlaminar shear strength properties of ecopoxy-based natural fiber hybrid biocomposites were investigated and the results are presented in the following graphs. For each sample, a total of five composites were tested and the dimensions were taken as 79.6 × 12.7 × 4.6 mm length, width, and thickness. Notch depth is taken as half of the specimen thickness.

Figure 6.4a, b shows the interlaminar shear strength properties of hemp-flax, noil hemp-noil flax natural fiber hybrid biocomposites. As it can be seen from the obtained results, the shear strength values of hybrid biocomposites are superior to the pure fiber (individual fiber) biocomposites in both cases. Natural hemp and flax fiber biocomposites showed better shear strength properties than noil fiber biocomposites. In the first set (H/F) of biocomposites 25H/15F hybrid biocomposites showed the

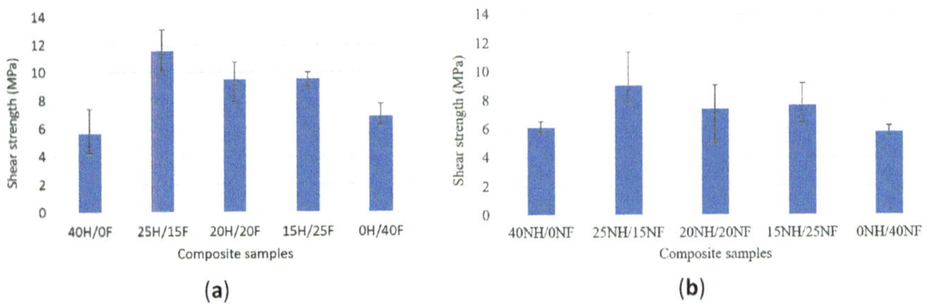

Fig. 6.4 Interlaminar shear strength behavior of **a** H/F and **b** NH/NF natural fiber hybrid biocomposites

highest shear strength as 13.10 MPa, and Pure hemp (40H/0F) showed the least as 5.67 MPa. Whereas in the second set (NH/NF) of biocomposites 25NH/15NF composites showed the highest as 9.03 MPa and pure noil flax (0NH/40NF) showed the least as 5.43 MPa. Overall hemp fiber-rich hybrid biocomposites showed superior properties in both cases.

Flexural strength test. The Flexural strength properties of ecopoxy-based natural fiber hybrid biocomposites were investigated and the results are presented in the following graphs. A 3-point bending test was performed on Tinus Olsen at 21° temperature and 28% of relative humidity to calculate the bending strength of biocomposites. For each sample, a total of five composites were tested and the dimensions were taken as 80 × 20 × 4.6 mm length, width, and thickness as per ASTM standards.

The flexural strength properties of natural fiber hybrid biocomposites are given in Fig. 6.5a, b. In both cases, pure composites showed the least flexural strength values whereas hybrid composites showed superior properties and there is minimal variation in the outcomes. Hemp/flax fiber hybrid biocomposites showed a range between 17 and 30 MPa whereas noil fiber biocomposites exhibited in between 13 and 24 MPa. This is owing to the fiber's hybridization effect, which gives the biocomposites higher flexural strength in both circumstances. Poor bonding between reinforcement and resin material in pure biocomposites, resulting in the agglomeration and hence a loss of strength, as well as fabrication faults, could be the cause of pure biocomposites reduced strength. The hybridization impact was observed clearly in both cases in terms of increasing strength.

Flexural modulus. The results presented in Fig. 6.6a, b show the assessment of flexural modulus properties of various hybrid fiber biocomposites. The results followed a similar trend to the previous results. The hemp and flax fiber hybrid biocomposites showed higher flexural modulus properties than noil hemp and noil flax fiber hybrid biocomposites. Pure noil hemp fiber biocomposites showed the least flexural modules as 0.261 GPa whereas 25H/15F showed the highest as 0.825 GPa. The is not much variation in hemp and flax hybrid composites and the values are close to each other, whereas in noil fiber biocomposites variation is quite observable.

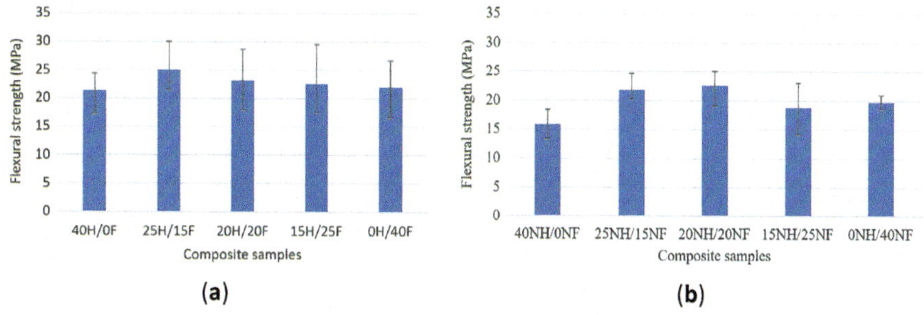

Fig. 6.5 Flexural strength behavior of **a** H/F and **b** NH/NF natural fiber hybrid biocomposites

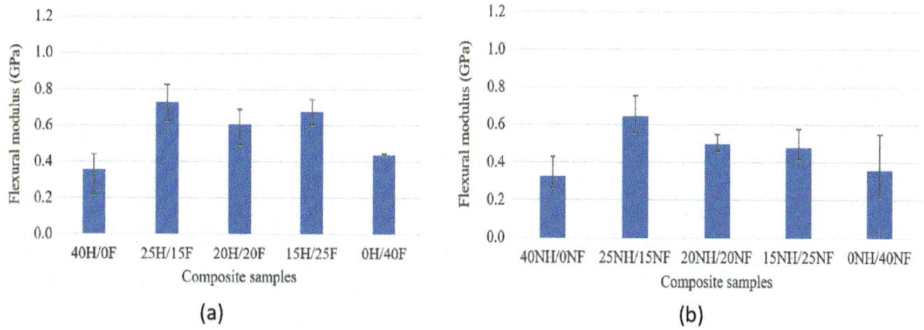

Fig. 6.6 Flexural modulus of **a** H/F and **b** NH/NF natural fiber hybrid biocomposites

Contact angle measurement analysis. The contact angle measurement illustrations of the biocomposites are shown in Fig. 6.7a, b. The lighting has been meticulously planned to eliminate any light reflections that may detract from the measurement. Precautions were also taken to keep the drips from becoming contaminated by air pollutants such as dust and particulates.

The graphical representation of measured contact angles for various natural fiber hybrid biocomposites is shown in Fig. 6.8a, b. From the obtained results the contact

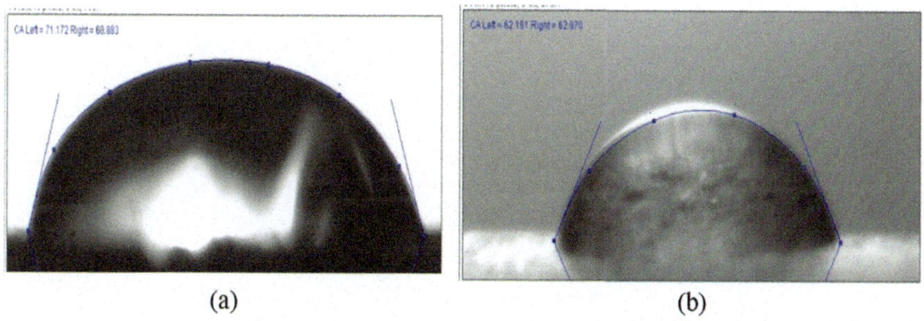

Fig. 6.7 Contact angle image illustration for **a** H/F and **b** NH/NF biocomposites

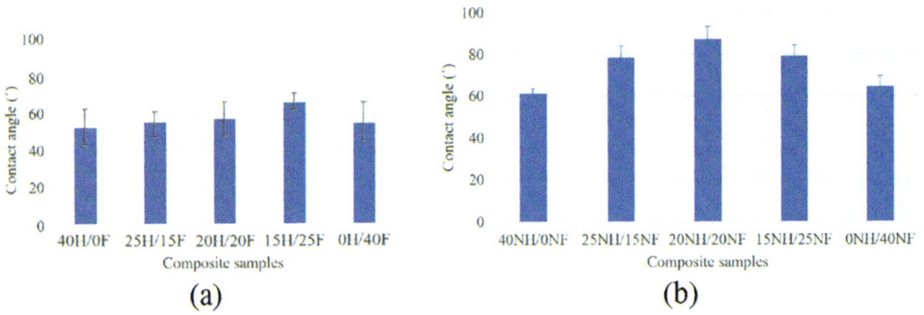

Fig. 6.8 Contact angle measurement for **a** H/F and **b** NH/NF natural fiber biocomposites

angle maximum was observed for 20NH/20NF hybrid composite which is 87.3° and the lowest is observed for 40H/0F which is 53°. Hemp and flax biocomposites showed the contact angle ranging from 53° to 66° and noil fiber composites exhibited in between 61° and 87° which means that all the composites showed hydrophilic surface properties. Whereas 20NH/20NF hybrid fiber composites are almost fallen in the range of hydrophobic material properties.

Fiber morphology studies. The fibers surface morphology and the effect of chemical treatment on natural fibers used in this research work are analyzed by using an optical microscope. The following figures differentiate the treated and untreated natural fibers.

The surface morphology of the fibers was studied by an optical microscope. The microscopic images of the fiber surfaces are presented in Figs. 6.9 and 6.10. Figure 6.9a, b show the optical microscope images of untreated hemp and flax fibers which have a smooth surface finish. This is due to the presence of adhesive polysaccharides on the surface of fibers. Whereas Fig. 6.10a, b shows the images of treated hemp and flax fibers which have a rough surface finish. This is due to the alkaline (NaOH) treatment of natural fibers. The polysaccharides are removed from the surface of the fibers and fibers are separated from each other. The roughness on the fibers improves the adhesion between reinforcement and matrix material.

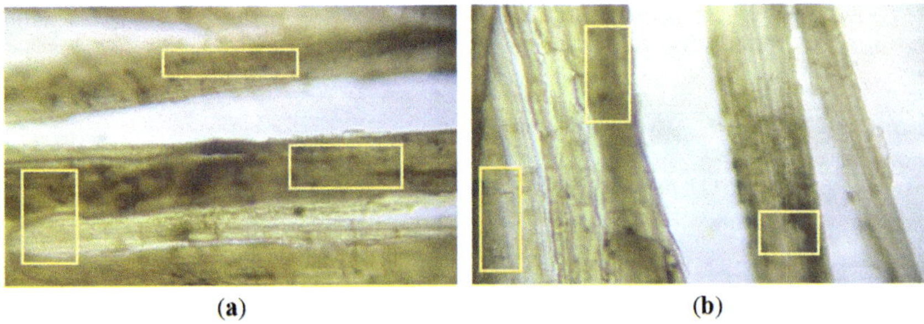

Fig. 6.9 Microscopic images for untreated **a** hemp and **b** flax fibers having smooth surface finish

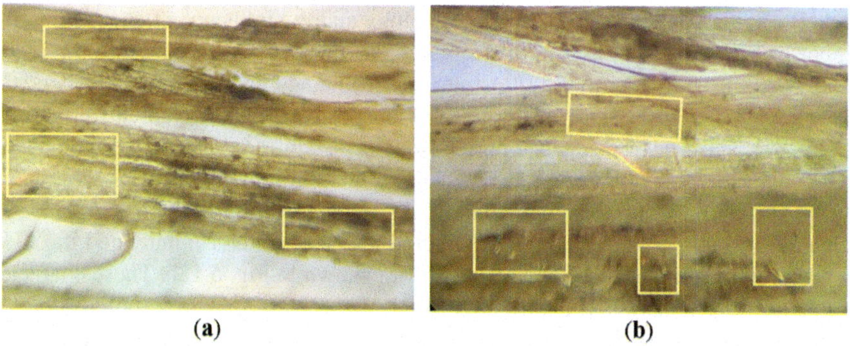

Fig. 6.10 Microscopic images for treated **a** hemp and **b** flax fibers having rough surface finish

The microcracks and kink bands are two forms of flaws that are normally predicted to be found in the cell walls of natural fibers. Figure 6.11a, b shows the microcracks observed in optical micrographs of noil hemp and noil flax fibers. Whereas Fig. 6.11c, d shows the microscopic images for treated noil hemp and noil flax fibers without microcracks in the fiber cell wall. The kink bands were observed in untreated noil hemp fibers in Fig. 6.11e. After chemical treatment, the kink bands were eliminated and shown in Fig. 6.11f. The reinforcement materials act as a load carrier in a composite material. So, the presence of microcracks and kink bands affects the fiber properties and lowers the strength. We can say that to avoid these kinds of defects in natural fibers chemical treatment is one of the best solutions.

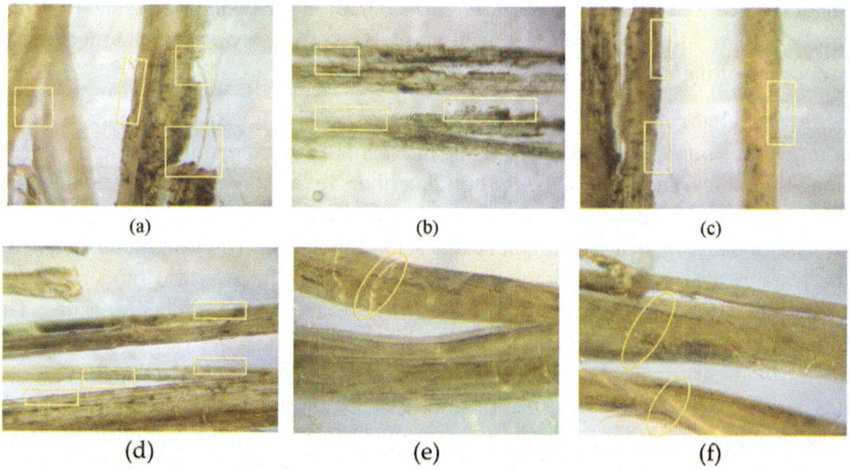

Fig. 6.11 Microscopic images for untreated noil hemp (**a**), noil flax (**b**) fibers having microcracks, treated noil hemp (**c**), and noil flax (**d**) fibers without microcracks, untreated noil hemp (**e**) with kink bands treated noil hemp (**f**) without kink bands

FTIR analysis. The fibers surface morphology and the influence of chemical treatment on natural fibers used in this research work are analyzed by using the FTIR analysis. The following figures illustrate the treated and untreated natural fibers.

The above Fig. 6.12a, b show the influence of chemical treatment on natural fibers was analyzed by using the Fourier Transformed infrared spectroscopy analysis. The character peaks of treated hemp (T–H) treated noil hemp (T–N–H), untreated hemp (U–H), and untreated noil hemp (U–N–H) fibers are shown in Fig. 6.12a. The character peaks of treated flax (T–F) treated noil flax (T–N–F), untreated flax (U–F), and untreated noil flax (U–N–F) fibers are shown in Fig. 6.12b. These character peaks of fiber materials are attributed to the presence of cellulose, hemicellulose, and lignin content in the fiber material. In general, the IR spectra for natural fibers are presented in the 3250–2650 cm $^{-1}$ range and also the large bands are indicated for the deformation of O–H groups. For the untreated hemp, flax, noil hemp, and noil flax fibers this peak is found in a range between 3374 and 2846 cm^{-1}. The detected absorption band is connected to the O–H group deformation. The second large peak followed by the O–H group is observed at 2672 cm^{-1} which exhibits the C–O group of the hemicellulose in UH. The characteristic peak at 1584 cm^{-1} is characteristic of the N–H group of cellulose. The peak at 523 cm^{-1} is related to the irregular deformation of C–I of the cellulose and hemicellulose. Whereas in treated fibers the highest peak was observed around 2780 cm^{-1}. The large peak represents the O–H groups in cellulose. The intensity of this peak is higher than the peak observed in the untreated fibers. It was observed from the figure that the obscene C–O band of the carboxyl group is present in the hemicellulose of natural fibers. Which indicates the removal of hemicellulose in the alkaline-treated natural fibers. Also, in the treated fibers the lignin traces were not absorbed much. So due to the elimination of hemicellulose and lignin content in the natural fibers the moisture intake of natural fibers can be reduced and which adversely affects the mechanical properties of natural fiber composites.

Scanning electron microscope (SEM) analysis. The fracture mechanism of the composites samples is analyzed by using the scanning electron microscope. The SEM images of hemp/flax and noil hemp/noil flax fibers hybrid biocomposites flexural test specimens were taken to analyze the Defects in them.

The above Fig. 6.13a–e depicts the scanning electron micrographs taken for the fracture flexural test specimens of hemp and flax fiber hybrid biocomposites. From Fig. 6.13a, e, the breakage of fibers was observed in pure biocomposites. The breakage of the fibers lowers the strength of the composites. Because reinforcements are the load-carrying agent in a composite material when it is subjected to the external loading(force). And the presence of microcracks and air voids can be observed in hybrid composites (Fig. 6.13b–d). the presence of defects in the composite material may lower the strength of the composites. The reasons for these defects can be fabrication error, residual stress during the curing process, and poor adhesion between matrix and reinforcement. Overall 40H/0F, 0H/40F biocomposites are having more defects than hybrid biocomposites it leads to weaking the strength. Figure 6.13f–i shows SEM micrographs of the shattered surfaces of the generated bio-composites

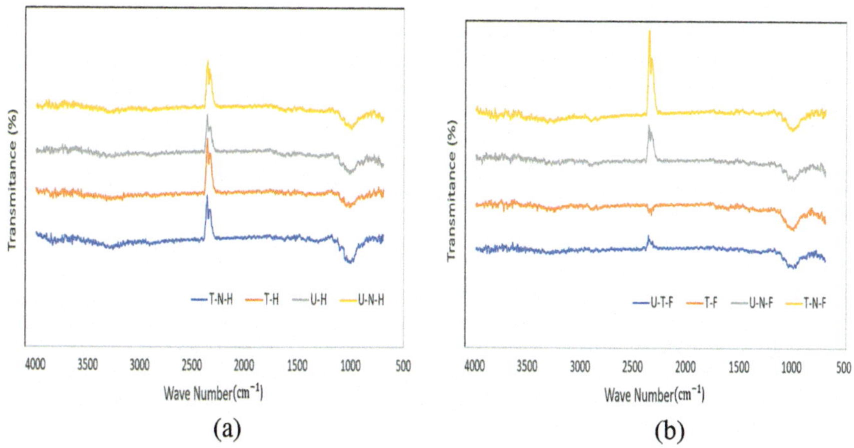

Fig. 6.12 FTIR analysis of treated and untreated hemp and noil hemp fibers (**a**) treated and untreated flax and noil flax fibers (**b**)

(NH/NF) coming from samples that perform well in most of the attributes studied. It was observed from the results that the presence of fiber clusters and twisted fibers in the pure noil biocomposites (Fig. 6.13f, j). This may cause the agglomeration of the fibers in the composite materials and lowers the strength as well. The presence of air pits is observed in Figs. 6.13g–i which attracts the moisture easily. The reasons for the presence of internal defects are explained earlier.

The rate of absorption for the composite material mainly depends on the presence of porosity content and fiber content in it. Figure 6.14a–e shows the presence of porosity (air voids) in hemp and flax fiber hybrid composites and (f)–(j) shows the noil hemp and noil flax fiber hybrid composites. It was observed that pure biocomposites have a larger void percentage whereas hybrid composites have a low percentage. The porosity content in 25H/15F (Fig. 6.14b) is minimal so it enhances that improvement in strength. Also, it was observed that 40NH/0NF composites have a higher void percentage and 25NH/15NF fiber composites have the least void content. In both cases hybrid biocomposites have a less void percentage, which reflects the improvement in strength whereas pure biocomposites have more void content. Hence the rate of absorption is high in pure composites and less in hybrid composites in both cases.

SEM–EDX spectrum analysis. The chemical composition presented in the natural fiber biocomposites are analyzed with scanning electron microscopic energy dispersive X-ray spectroscopy and the results are shown in the following figures.

The Chemical composition presented in the hemp and flax fiber hybrid biocomposites at surface position is studied by using the SEM-Energy dispersive X-Ray spectrometer and the results are described in Fig. 6.15a, b. The percentage of weight concentration and atomic concentration of the individual elements at the surface position are given in Table 6.3. From the results, it was observed that there is no such variation in the biocomposites at the surface position. In both fibers, the percentage

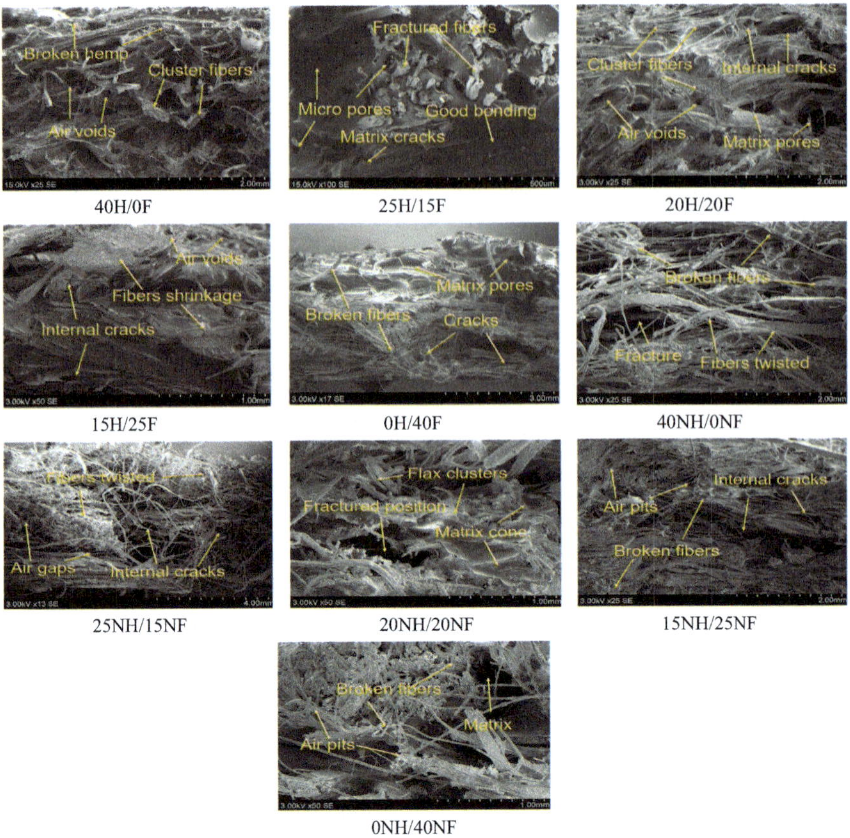

Fig. 6.13 SEM micrographs for hemp/flax fiber biocomposites flexure test fracture samples

of carbon takes the top place followed by oxygen. The presence of sodium and silicon is observed in hemp fiber biocomposites though the percentage is minimal. It was not observed in the flax fiber composites. The presence of oxygen in the composite surface may form the hydroxyl bonds with hydrogen presented in the cellulose content, which lead to attract the moisture content and weaken the composite strength.

After the surface position, the spectrum analysis was carried out at the fractured position to understand the internal composition of hemp and flax fiber biocomposites. The resultant micrographs are shown in the following figures.

The Chemical composition presented in the hemp and flax fiber hybrid biocomposites at fractured position is studied by using the SEM–EDX spectrometer and the results are described in Fig. 6.16a, b. The percentage of weight concentration and atomic concentration of the individual elements at the fractured position are given in Table 6.4. From the results, it was observed that there is some variation in the biocomposites at the fractured position unlike at the surface. In both fibers, the

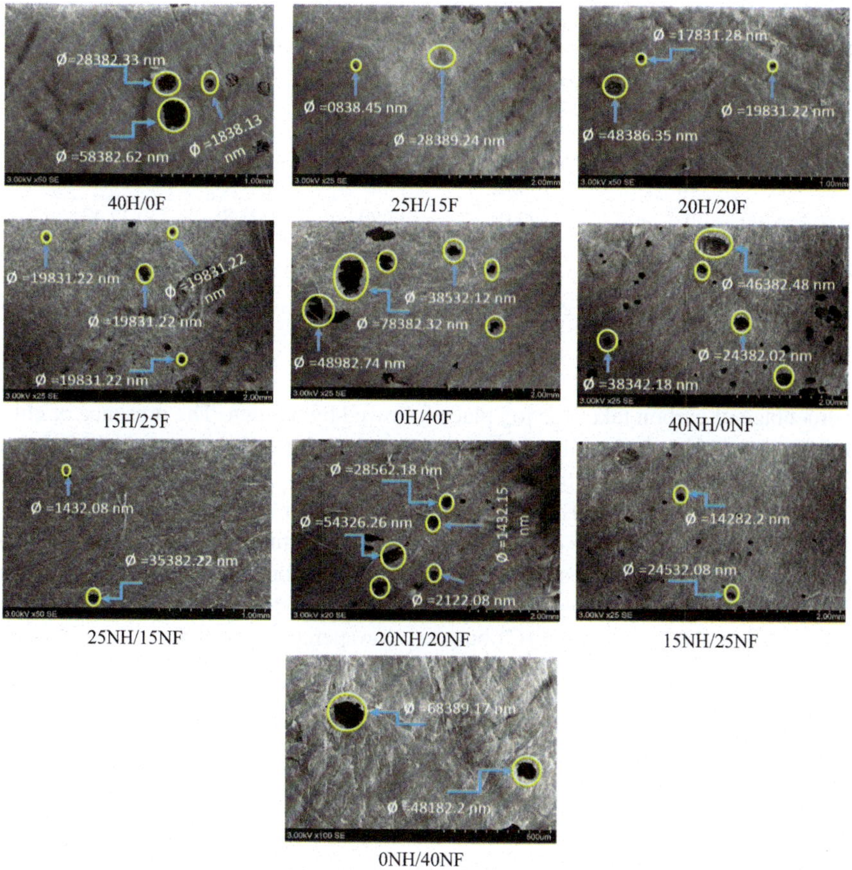

Fig. 6.14 SEM micrographs for porosity content in Hemp and flax fiber hybrid biocomposites

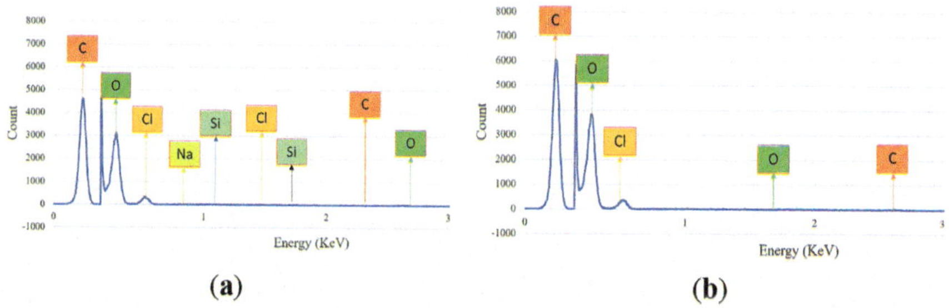

Fig. 6.15 Spectrum analysis of chemical composition for **a** hemp and **b** flax fiber biocomposites at surface position

Table 6.3 Spectrum analysis for chemical configuration of natural fiber hybrid biocomposites at the surface position

Element type	Atomic weight	Hemp biocomposite		Flax biocomposite	
		Weight (%)	Atomic conc. (%)	Weight (%)	Atomic conc. (%)
Carbon (C)	12	76.29	81.61	77.17	82.18
Oxygen (O)	16	22.04	17.70	21.85	17.47
Chlorine (Cl)	35.5	1.03	0.37	0.97	0.35
Sodium (Na)	23	0.28	0.16	–	–
Silicon (Si)	28	0.35	0.16	–	–

percentage of carbon takes the top place followed by oxygen. The presence of chlorine is observed in hemp fiber biocomposites though the percentage is very small. But it was not observed in the flax composites.

After the spectrum acquisition for hemp and flax fiber biocomposites at the surface and fractured positions, we can deduce that the proportion of oxygen is more at the fractured place than at the surface and it is opposite for carbon percentage. And it was also observed that the presence of chlorine (Cl), sodium (Na), silicon (Si) along with oxygen (O), and carbon (C) but very low percentages. This reflects that the

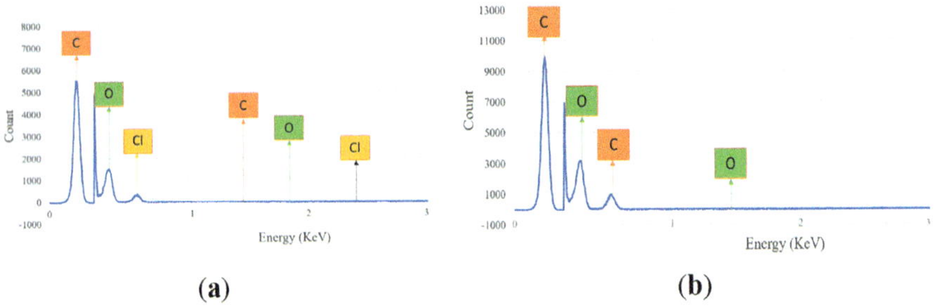

Fig. 6.16 Spectrum analysis of chemical structure for **a** hemp and **b** flax fiber biocomposites at fractured position

Table 6.4 Spectrum analysis for chemical configuration of natural fiber hybrid biocomposites at the fracture position

Element type	Atomic weight	Hemp biocomposite		Flax biocomposite	
		Weight (%)	Atomic conc. (%)	Weight (%)	Atomic conc. (%)
Carbon (C)	12	63.73	70.26	68.82	74.62
Oxygen (O)	16	35.67	29.52	31.18	25.38
Chlorine (Cl)	35.5	0.60	0.22	–	–

concentration matrix (ecopoxy) material is more on the surface than the fracture position. Which leads to a smooth surface finish at the surface of the composite.

6.4 Conclusions

In the present study, natural fiber-reinforced ecopoxy hybrid biocomposites comprising hemp, flax, noil hemp, and noil flax were fabricated employing the hand layup technique, followed by a compression molding process. Prior to fabrication, the fibers were subjected to treatment with a 5% NaOH solution. The resulting composites were subsequently subjected to mechanical and morphological testing, with the outcomes reported as follows.

Hemp and flax fiber hybrid biocomposites showed superior mechanical properties than noil hemp and noil flax fiber biocomposites. In both cases, hybrid biocomposites showed improved properties than pure biocomposites. From interlaminar shear test results, hemp and flax fiber biocomposites showed a range between 5.67 and 13.10 MPa whereas noil fibers as 4.9–11.3 MPa. From flexural test results, 25H/15F hybrid biocomposites exhibited the maximum flexural strength and modulus as 30.04 MPa and 0.825 GPa and pure noil hemp (40NH/0NF) fiber biocomposites exhibited the minimum as 13.31 MPa and 3.95 GPa.

All biocomposites demonstrated contact angles of <90° based on contact angle measurements, signifying their hydrophilic surface characteristics. However, the noil fiber biocomposites exhibited contact angles approaching 90°, suggesting a rough surface finish and a tendency towards hydrophobic properties. Fiber morphology studies on both treated and untreated fibers were conducted utilizing an optical microscope. The results were delineated by contrasting the defects present in both treated and untreated fibers. The fracture mechanisms and porosity content within the biocomposites were examined using a scanning electron microscope. The chemical composition spectra of hemp and flax fiber biocomposites were acquired using SEM–EDX spectrographs. Overall, the hemp and flax fiber hybrid biocomposites exhibited superior properties compared to noil fiber biocomposites, thereby suggesting their efficacy as reinforcements in applications such as lightweight structures, biosensor chips, flexible electronics, microfluidics, and bio-composite applications.

References

1. Atmakuri A, Palevicius A, Janusas G, Eimontas J (2022) Investigation of hemp and flax fiber-reinforced EcoPoxy matrix biocomposites: morphological, mechanical, and hydrophilic properties. Polym Basel MDPI 14(21):1–21. https://doi.org/10.3390/polym14214530
2. Faruk O, Bledzki AK, Fink H, Sain M (2012) Biocomposites reinforced with natural fibers: 2000–2010. Prog Polym Sci 37:1552–1596

3. Sanyang ML, Sapuan SM, Jawaid M, Ishak MR, Sahari J (2016) Recent developments in sugar palm (*Arenga pinnata*) based biocomposites and their potential industrial applications: a review. Renew Sustain Energy Rev 54:533–549
4. Etaati A, Pather S, Rahman M, Wang H (2015) Ground hemp fibers as filler/reinforcement for thermoplastic biocomposites. Adv Mater Sci Eng 11:1–12
5. Etaati A, Pather S, Fang Z, Wang H (2014) The study of fibre/matrix bond strength in short hemp polypropylene composites from dynamic mechanical analysis. Compos B Eng 62:19–28
6. da Luz FS, Filho F, Da CG, Oliveira MS, Cassiano Nascimento LF, Neves Monterio S (2020) Composites with natural fibers and conventional materials applied in a hard armor: a comparison. Polymers 12(9):1920
7. Peças P, Carvalho H, Salman H, Leite M (2018) Natural fibre composites and their applications: a review. J Compos Sci. 2(4):66
8. Ramakrishnan G, Ramnath BV, Elanchezhian C, Arun Kumar A, Gowtham S (2019) Investigation of mechanical behaviour of basalt-Banana hybrid composites. SILICON 11:1939–1948
9. Sair S, Oushabi A, Kammouni A, Tanane O, Abboud Y, Bouari AEL (2018) Mechanical and thermal conductivity properties of hemp fiber reinforced polyurethane composites. Case Stud Constr Mater 8:203–212
10. Moonart U, Utara S (2019) Effect of surface treatments and filler loading on the properties of hemp fiber/natural rubber composites. Cellulose 26:7271–7295
11. Moudood A, Rahman A, Öchsner A, Islam M, Francucci G (2019) Flax fiber and its composites: an overview of water and moisture absorption impact on their performance. J Reinf Plast Compos 38:323–339
12. Yan L, Chouw N, Jayaraman K (2014) Flax fibre and its composites: a review. Compos B Eng 56:296–317
13. Sinha AK, Narang HK, Bhattacharya S (2017) Mechanical properties of natural fibre polymer composites. J Polym Eng 37:879–895
14. Rokbi M, Osmani H, Imad A, Benseddiq N (2011) Effect of chemical treatment on flexure properties of natural fiber-reinforced polyester composite. Proced Eng 10:2092–2097
15. Neto JSS, Lima RAA, Cavalcanti DKK, Souza JPB, Agular RAA, Banea MD (2019) Effect of chemical treatment on the thermal properties of hybrid natural fiber-reinforced composites. J Appl Polym Sci 136:47154
16. Shuhimi FF, Abdollah MF, Bin KMA, Masjuki HH, Sushella Edayu Mat Kamal AM, Amiruddin H (2017) Effect of operating parameters and chemical treatment on the tribological performance of natural fiber composites: a review. Sci Technol 35:512–524
17. Pereira AL, Banea MD, Neto JSS, Cavalcanti DKK (2020) Mechanical and thermal characterization of natural intralaminar hybrid composites based on sisal. Polymers 12:866
18. Cavalcanti DKK, Banea MD, Neto JSS, Lima RAA, da Silva LFM, Carbas RJC (2019) Mechanical characterization of intralaminar natural fibre-reinforced hybrid composites. Compos B Eng 175:107149
19. Ravishankar B, Nayak SK, Kader MA (2019) Hybrid composites for automotive applications: a review. J Reinf Plast Compos 38(18):835–845
20. Yahaya R, Sapuan SM, Jawaid M, Leman Z, Jainudin ES (2016) Effect of fibre orientations on the mechanical properties of kenaf–aramid hybrid composites for spall-liner application. Def Technol 12:52–58
21. Ashori A (2010) Hybrid composites from waste materials. J Polym Environ 18:65–70
22. Ashori A, Nourbakhsh A (2008) A comparative study on mechanical properties and water absorption behavior of fiber-reinforced polypropylene composites prepared by OCC fiber and aspen fiber. Polym Compos 29(5):574–578
23. Sabuncuoglu B, Cakmakci O, Kadioglu FS (2020) Fiber/matrix interface stress analysis of flax-fiber composites under transverse loading considering material nonlinearity. J Reinf Plast Compos 39:345–360
24. Jaafar J, Siregar JP, Mohd Salleh S, Mohd Hamdan MH, Cionita T, Rihayat T (2019) Important considerations in manufacturing of natural fiber composites: a review. Int J Precis Eng Manuf Technol 6:647–664

25. Pickering KL, Efendy MGA, Le TM (2016) A review of recent developments in natural fibre composites and their mechanical performance. Compos A Appl Sci Manuf 83:98–112
26. Supian ABM, Jawaid M, Rashid B, Fouad H, Saba N, Hom N, Dhakal N, Khiari R (2021) Mechanical and physical performance of date palm/bamboo fibre reinforced epoxy hybrid composites. J Mater Res Technol 15:1330–1341
27. Afzaluddin A, Jawaid M, Salit MS, Ishak MR (2019) Physical and mechanical properties of sugar palm/glass fiber reinforced thermoplastic polyurethane hybrid composites. J Mater Res Technol 8:950–959
28. Ramesh M, Dhilip JDJ, Akilan S, Ganesh Prasad B, Hari Shankar G (2020) Influence of stacking sequence on mechanical and thermal characteristics of banana-pineapple fiber reinforced epoxy composites. J Nat Fibers 17:1–14
29. Zalinawati M, Siregar JP, Tezara C, Sazali N, Jaafar J, Oumer AN, Hamdan MHM (2020) The effect of fibre treatment on water absorption and mechanical properties of buri palm (*Corypha utan*) fibre reinforced epoxy composites. J Mech Eng Sci 14(4):7379–7388
30. Torres-Arellano M, Renteria-Rodríguez V, Franco-Urquiza E (2020) Mechanical properties of natural-fiber-reinforced biobased epoxy resins manufactured by resin infusion process. Polymers 12(12):2841
31. Jamshaid H, Mishra R, Basra S, Rajput AW, Hasan T, Petru M, Choteborsky R, Muller M (2020) Lignocellulosic natural fiber reinforced bisphenol F epoxy based bio-composites: characterization of mechanical electrical performance. J Nat Fibers. 19(9):1–10
32. Rajak D, Pagar D, Menezes P, Linul E (2019) Fiber-reinforced polymer composites: manufacturing, properties, and applications. Polymers 11:1667
33. Terwadkar AA, Walame MV (2019) Mechanical properties of banana fabric and kenaf fiber reinforced epoxy composites: effect of treatment and hybridization. Int J Innov Technol Explor Eng 8:1870–1874
34. Neves ACC, Rohen LA, Mantovani DP, Carvatho PRG, Vieira CMF, Lopes FPD, Simonassi NT, da Luz FS, Monteiro SN (2020) Comparative mechanical properties between biocomposites of epoxy and polyester matrices reinforced by hemp fiber. J Mater Res Technol 9:1296–1304
35. Thiagamani SMK, Krishnasamy S, Muthukumar C, Tengsuthiwat J, Nagarajan R, Siengchin S, Ismail SO (2019) Investigation into mechanical, absorption and swelling behaviour of hemp/sisal fibre reinforced bioepoxy hybrid composites: effects of stacking sequences. Int J Biol Macromol 140:637–646
36. Bazan P, Mierzwiński D, Bogucki R, Kuciel S (2020) Bio-based polyethylene composites with natural fiber: mechanical, thermal, and ageing properties. Materials 13:2595
37. Mochane MJ, Mokhena TC, Mokhothu TH, Mtibe A, Sadiki ER, Ray SS, Ibrahim ID, Daramola OO (2019) Recent progress on natural fiber hybrid composites for advanced applications: a review. Exp Polym Lett 13(2):159–198
38. Ramu P, Jaya Kumar CV, Palanikumar K (2019) Mechanical characteristics and terminological behavior study on natural fiber nano reinforced polymer composite: a review. Mater Today Proc 16:1287–1296
39. Saba N, Tahir P, Jawaid M (2014) A review on potentiality of nano filler/natural fiber filled polymer hybrid composites. Polymers 6:2247–2273
40. Ku H, Wang H, Pattarachaiyakoop N, Trada M (2011) A review on the tensile properties of natural fiber reinforced polymer composites. Compos B Eng 42:856–873

Chapter 7
Basalt/E-Glass Woven Fabric–Reinforced Composites

Abstract This chapter elucidates the development of innovative hybrid composites incorporating filler particles, produced via a proposed novel fabrication technique (Atmakuri et al. in Micromach Basel MDPI 13:1–14, 2022 [1]). The hybrid composites were synthesized using a Basalt and E-glass woven fabric reinforced with an epoxy resin matrix amalgamated with graphite powder nanoparticles. Six sets of samples were crafted employing the Vacuum-assisted free lamination compression molding technique. Subsequent to fabrication, the wettability, mechanical properties—comprising tensile, flexural, and impact properties—and moisture characteristics were assessed. The surface morphology and chemical composition of the composite samples were scrutinized utilizing a Scanning Electron Microscope (SEM) and spectroscopy. The findings revealed that integrating filler materials within hybrid composites enhances their properties. Basalt/E-glass hybrid composites containing 10% graphite material demonstrated superior mechanical properties relative to other composites, exhibiting high-quality, enhanced adhesion, and improved surface morphology. Consequently, these novel composites, endowed with an amalgamation of exceptional attributes, hold potential for integration in the design of flexible electronics and microfluidic devices as a structural layer of the system. The pronounced flexibility and favorable surface tension of the designed composites render them desirable for application within the thermal imprint technique employed in microfluidic channel design.

7.1 Introduction

The significance of natural fibers in the advancement of fiber-reinforced hybrid polymer composites is progressively acknowledged in contemporary times, as it is intrinsically linked to environmental considerations and the principles of a circular economy [2]. The integration of natural fibers with synthetic fibers within reinforced hybrid polymer composites presents numerous advantageous characteristics, including reduced weight, minimized lifecycle expenses, high adhesion qualities, flexibility, and enhanced mechanical properties [3]. Nature provides a substantial

A. Atmakuri et al., *Fiber-Based Hybrid Composites for Engineering*, Springer Series on Polymer and Composite Materials, https://doi.org/10.1007/978-3-031-93144-4_7

source for the production of composite materials. Examples of regular fiber composites include Hemp, Flax, Sisal, Palm, Okra, Banana, Wood, Bamboo, Jute, Sugarcane, Cowhide, among others. The trunks of palm trees serve as authentic examples of sandwich constructions within various composite forms [4]. Evaluations of static and dynamic mechanical properties are fundamental components of composites and are receiving significant attention in their application across diverse domains [5].

In recent years, synthetic fiber-based composites have emerged as the foremost advanced composite materials utilized in automotive, construction, marine, sporting goods, biomedical, microfluidics, and various engineering applications, attributable to their accessibility, high strength, and modulus, coupled with excellent corrosion and fatigue resistance properties [6]. While natural fibers present certain advantages within composite materials, they also pose disadvantages such as fiber shrinkage, poor adhesion, and moisture absorption, which may adversely affect the mechanical properties. Additionally, the inherent brittleness of most synthetic fibers poses a risk of catastrophic failure in composite materials [7]. Synthetic fibers, including carbon, glass, mica, and nylon fibers, significantly contribute to the composite industry. Fiber-reinforced composite materials exhibit a superior combination of strength and modulus compared to many traditional metallic materials. A fiber-reinforced composite is characterized by low specific gravity, strength-to-weight ratio, and modulus-to-weight ratios, which surpass those of metallic materials [8, 9]. The exhaustion strength-to-weight ratio influences the fatigue performance of composite materials. Natural fibers, being lighter than metallic fibers, are advantageous in numerous weight-sensitive applications such as aerospace, civil engineering, and the automotive sector [10].

Numerous researchers have shown a keen interest in the development of hybrid composites. This interest can primarily be ascribed to the detrimental environmental effects, elevated costs, and certain adverse mechanical properties associated with polymer composites [11]. Hybridization of polymer composites has been recognized for enhancing the mechanical properties beyond those of individual fiber composites. This is achieved through the integration of two or more reinforcing materials within a common matrix, which could be either Thermoset or Thermoplastic. The potential combinations include Natural fiber-Natural fiber, Natural fiber-Synthetic fiber, and Synthetic fiber-Synthetic fiber hybrid composites [12, 13]. These hybrid composites have been empirically demonstrated to mitigate the limitations inherent in individual fiber composites. Consequently, when optimally configured, hybrid composites exhibit superior quality and robust mechanical properties such as strength, stiffness, flame retardancy, and resistance to corrosion [14].

Among the prominent methodologies for comprehending the wettability characteristics of composite materials, a range of droplet-based techniques are employed [15], encompassing gravitational [16, 17], acoustic [18], and magnetic [19] driving methods. Of these droplet driving methodologies, contact angle measurement is particularly favored due to its precision, cost-effectiveness, and procedural simplicity. The investigation of the wettability of composite materials enables the elucidation of the material's surface attributes, such as hydrophilic or hydrophobic tendencies, and facilitates the understanding of the surface tension of driven droplets.

Consequently, the mechanical properties of thermoset and thermoplastic resin-based hybrid polymer composites, incorporating synthetic fibers such as carbon, nylon, glass, rayon, spandex, and acrylic, as well as natural fibers like hemp, flax, sisal, jute, bamboo, and palm, have been extensively explored by numerous researchers. Several studies have asserted that hybrid composites exhibit superior attributes in comparison to individual fiber composites. The static and dynamic mechanical properties of basalt fiber-reinforced polymer composites demonstrate notable sensitivity to strain rate, with mechanical properties such as tensile strength, elastic modulus, and failure strain exhibiting rapid enhancements when the strain rate exceeds 120s-1 [20]. Thus, replacing glass fiber with basalt fiber in the composites, it enhances mechanical properties by 35–42%, i.e., compressive strength, impact energy, and young's modulus [21]. And replacing glass fiber with E-glass fiber composites gives superior tensile strength [21, 22]. Moreover, applying plasma treatment on basalt fibers during the fabrication, the adhesion properties and surface roughness may be improved compared to those of commercial fibers [22]. The influence of waste basalt powder on rigid polyurethane foams had some influence, too. It was observed that the addition of basalt powder triggered an increase in the reactivity of the polyurethane system during the foaming process, but it caused a reduction of mechanical properties [23]. Also, the basalt powder can be used as an eco-friendly modifier for the epoxy resin improving the thermo-mechanical properties for an increase in filler content [24]. The mechanical properties of composites reinforced with fabrics having the same areal density confirmed that Basalt Fiber composites possess an elastic modulus higher than the corresponding glass fiber composites, while their tensile strength approaches that of equivalent Carbon fiber composites. And the results of fatigue behavior confirmed the better performances of Basalt fiber composites than the corresponding glass fiber composites, with higher stiffness retention at low fatigue loads, and better damping properties [25]. Thus, Basalt and hybrid laminates with an added configuration exhibited higher impact energy absorption capacity than glass fiber composites applying the hybridization of reinforcement material [26–32]. However, the lack of filler material in composite fabrication may improve adhesion but significantly reduce its mechanical properties. To overcome these problems an attempt was made in the present chapter.

IIn this chapter, epoxy polymer composites reinforced with Basalt fiber (BF) and E-glass fiber (EF) and incorporating graphite filler material at various weight fractions (5, 10, and 15%) were fabricated using the Vacuum-Assisted Free Lamination Compression Molding technique. Subsequently, the samples underwent a heat treatment process and were evaluated for their mechanical properties. Mechanical characterization was conducted through tensile, flexural, impact, hardness, and moisture analyses. The wettability of the composite materials was assessed through contact angle measurement, which facilitated an understanding of the material's surface properties, including its hydrophilic or hydrophobic nature, and the surface tension of applied droplets. Surface morphology and chemical composition were analyzed utilizing Scanning Electron Microscopy (SEM) and spectroscopy methods. The findings confirmed the significance of appropriately configured hybrid composites, demonstrating high-quality mechanical and adhesion properties as well as qualitative surface morphology. The results established the potential application of novel composites in the engineering of microelectromechanical devices, flexible electronics, and microfluidics.

7.2 Materials and Methods

Basalt (B) and E-glass (E) woven fabrics were used as a reinforcement, and graphite (G) powder nanoparticles were used as a filler material. Epoxy resin along with hardener was used as a matrix material. Both, Basalt and E-glass fabrics have the same areal density of 198 g/m^2 and the fabrics were taken in a square shape (300 mm × 300 mm) to fabricate the composites. The mechanical properties and chemical characteristics of the materials were considered before the experimentation in order to understand nature and behavior during the process (Table 7.1).

A total of six sets of hybrid composites were fabricated using the Vacuum-assisted free lamination compression molding technique. In each set of composites, a total of six sheets were taken to fabricate 3 mm thickness composite plates. To make the hybrid composites with filler material, the graphite powder nanoparticles with varying weight fractions (5, 10, and 15%) were mixed in epoxy resin for 20 min and then applied to the composite panels. The ply sequence and orientation for each composite described in Table 7.2.

After degassing, a nominal pressure of 1.5 MPa was applied, and the composite panels were cured under the same pressure and vacuum for 8 h at room temperature. Then, the composite panels were taken for a heat treatment process for eliminating the moisture content, which also serves as a post-curing process for composites. The heat treatment process was carried out for 8 h at 90 °**C** temperature. A total of six sets in each five samples were developed using different materials and filler concentrations. Thus, the composite notations are considered as pure Basalt (B), pure E-glass (E), Basalt and E-glass (B/E), Basalt and E-glass with 5% filler (B/E/G5), Basalt and E-glass with 10% filler (B/E/G10), Basalt and E-glass with 15% filler material (B/E/G15), respectively (Table 7.2).

Table 7.1 The mechanical properties of materials [33–36]

Property	Basalt (B)	E-glass (E)	Epoxy resin
Density (g/cm^3)	2.40–2.72	2.48–2.62	1.15–1.18
Diameter (μm)	15	6	–
Tensile strength (MPa)	4800	3200	72
Young's modulus (GPa)	90	70	2.9
Elongation (%)	3.15	4.0	5.3
Maximum temp. (°C)	645	450	–
Color	Brown	White	Pale

Table 7.2 Ply sequence and orientation of Basalt and E-glass hybrid composites

Composite	No. of Plys		% Of graphite filler material	Ply sequence	Orientation (°)
	B	E			
B	6	0	0	B–B–B–B–B–B	0–90–0–90–0–90
E	0	6	0	E–E–E–E–E–E	0–90–0–90–0–90
B/E	3	3	0	B–E–B–E–B–E	0–90–0–90–0–90
B/E/G5	3	3	5	B–E–B–E–B–E	0–90–0–90–0–90
B/E/G10	3	3	10	B–E–B–B–E–E	0–90–0–90–0–90
B/E/G15	3	3	15	B–E–B–E–B–E	0–90–0–90–0–90

Analytical equipment. After the fabrication process, composites were allowed for wettability test and mechanical characterization by conducting various tests. In each set of composite groups, a total of five samples were tested according to ASTM standards (30 measurements in total). Thus, performed mechanical tests were the flexural test, tensile test, impact test, and moisture analysis. The morphological studies were carried out using the Scanning Electron Microscope (SEM) and the chemical compositions of the composite samples were tested using the SEM–EDX spectrometer.

Since fabricated hybrid composites contain the filler material throughout the surface and in the future researched are intended to use for flexible electronics and microfluidic devices, the wettability property of the fabricated composite surface plays a crucial role. The same test was conducted on composites without filler material for comparison purposes, as well. The wettability properties of novel composite materials were measured using the contact angle measurement technique (Fig. 7.1).

The experiment was carried out in a dim laboratory space with ambient lighting. It was vital for the measurement accuracy and image analysis that the liquid drop look black, thus the light settings were set up accordingly. The experiment was set up to eliminate any light reflections that would have tampered with the measurement (Fig. 7.1). Additionally, precautions were taken to keep airborne contaminants like dust and particulates from contaminating the drips. Also, with the critical distances between the parts were kept constant during the measurements (Fig. 7.2).

Further, flexural tests were performed on fabricated samples to analyze the flexural strength and modulus, i.e. a three-point bowing tests were performed on each set of

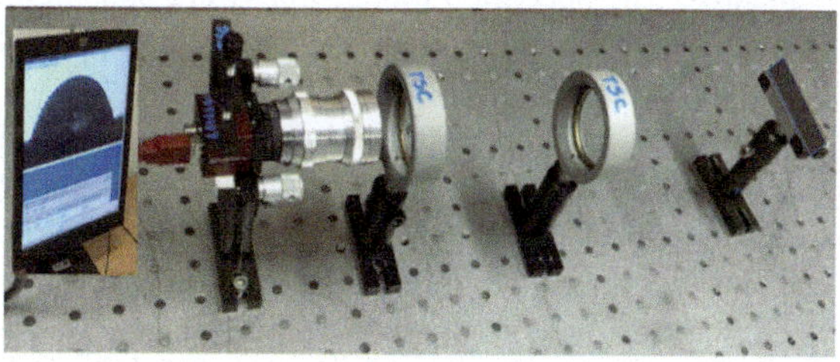

Fig. 7.1 A contact angle measurement setup

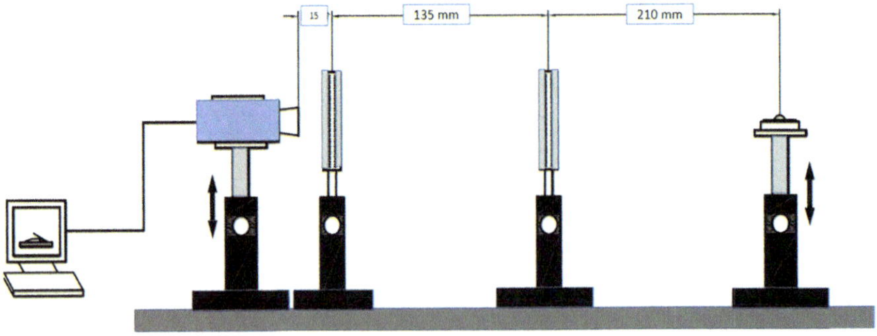

Fig. 7.2 Critical distance between contact angle testing device

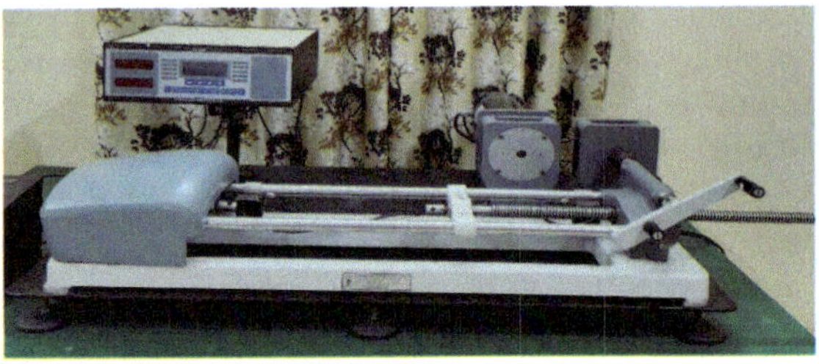

Fig. 7.3 Electronic tensiometer for flexural and tensile properties testing

composites according to ASTM D-790M-86 testing standards (total of 30 measurements). The dimensions of the samples were considered as 100 mm × 25 mm × 3 mm (length, width, thickness). The three-point bowing tests were conducted on an electronic tensiometer (METM 2000 ER-1 model (Plate II-18), provided by M/S Microtech Pune, India), with outside rollers in diameter of 70 mm. Thus, composites were loaded at a strain rate of 0.2 mm/min (Fig. 7.3).

Tensile tests were performed on the same electronic tensiometer which was used for flexural tests, at a crosshead speed of 0.2 mm/min. An electronic micrometer was utilized to quantify the necessary thickness and width of composite examples. The check length, width, and thickness were estimated with a 0.001 mm negligible count mechanized micrometer. This electronic tensiometer was fixed with weight and expansion pointers, which had an insignificant count of 0.01 kg and 0.01 mm, independently. The tensile tests were performed according to ASTM-D3039 testing standards. The specimen dimensions were considered 165 mm × 20 mm × 3 mm (length, width, thickness). In each set of composites, a total of five samples were tested (total 30 measurements).

Impact tests were performed on a falling dart impact testing machine (Impact analyzer M/S International equipment, Mumbai, India). The composite samples were tested as per ASTM D256-97 testing standards and the specimen dimensions were 63.5 mm long, 12.36 mm wide, and 3 mm in thickness. Designed composites were fixed in the vertical position and a pendulum was used to break the composite. The pendulum sled was delivered from the locking position which is at a state of 1500 mm and hits the sample with a striking rate of 2.46 m/s. The sample was stripped, and energy was exhibited in joules by the pointer on the specific scale.

The amount of moisture absorbed by the composite samples when exposed to moisture, was analyzed using the moisture analysis test. The tests were performed according to ASTM-D 570 testing standards. The sample specifications were considered as 60 mm long, 20 mm wide, and 3 mm in thickness. The samples were placed in a container full of water at room temperature and the amount of moisture gained by the composites was measured for six weeks duration (readings taken after every 7 days).

Finally, morphological studies such as surface morphology, porosity content, and fracture mechanism were investigated using SEM and the chemical composition of the composite samples was analyzed by using SEM–EDX spectroscopy.

7.3 Results and Discussion

The Basalt and E-glass fiber hybrid composites (B, E, B/E, B/E/G5, B/E/G10, and B/E/G15) in the presence of graphite filler (with varying weight fractions 5, 10, and 15%) were tested to evaluate their mechanical characteristics, wettability and surface morphology.

Wettability analysis. The wettability properties of composite materials were investigated using contact angle measurement technique imaging water droplet images on the composite surface (Fig. 7.4). The volume of the water droplet was kept constant and the values are noted from the left lower endpoint to the right lower endpoint.

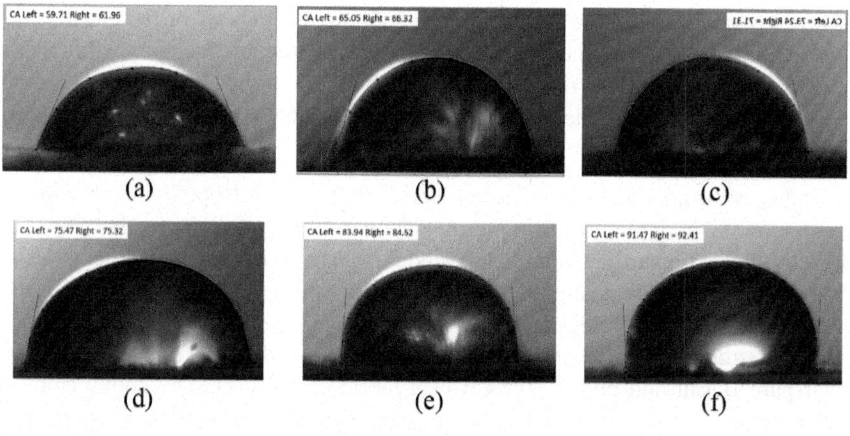

Fig. 7.4 The droplet profile's polynomial fit was achieved by using knots that go from the left to the right while accounting for the contact angle **a** B, **b** E, **c** B/E, **d** B/E/G5, **e** B/E/G10, and **f** B/E/G15

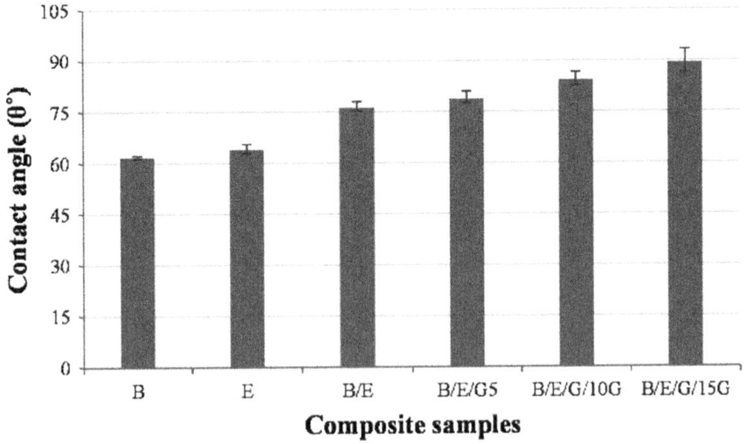

Fig. 7.5 Contact angle measurement for composite samples

Results (Fig. 7.5) showed the assessment of contact angle measurement of Basalt and E-glass woven fabric hybrid composites and that all samples surfaces exhibited hydrophilic properties. The measured contact angle formed by the water droplet on the B/E/G15 hybrid composite was 86.45° with an error of ±2.36°, being the maximum among all, whereas individual Basalt (B) composites the least contact angle of 61.76° with an error of ±1.84°, was exhibited. Thus, if the contact angle is <90° then it is considered hydrophilic, if it is more than 90° then it is hydrophobic [37–39]. Moreover, hybrid composites with graphite filler material tended to exhibit hydrophobic surface properties. To extend the analysis, the same tests were conducted varying the measurements in time. And results showed, that the contact angle decreased with an increase of time (at the continuous intervals of time such as 10, 20, 30 s) in all the cases due to seeping of a water droplet on the composite surface. The hydrophilic surface properties of the composite material plays a prominent role during the formation of micro channels on solid surfaces (composite material).

Flexural properties. The flexural properties, such as flexural strength and flexural modulus, of Basalt and E-glass woven fabric reinforced with epoxy matrix composites, including filler material (graphite powder particles) were investigated (Fig. 7.6). Results showed that, the strength and modulus were superior in hybrid composites compared to pure composites. The flexural properties increased with filler material and the impact of hybridization was observed, i.e. there was a slight increment in both, strength and flexural modulus in Basalt and E-glass hybrid composites over pure Basalt and pure E-glass woven fabric composites. The epoxy composites compared to pure Basalt (B) composites exhibited the least flexural strength and modulus at 27.82 MPa and 2.06 GPa, respectively. Thus, Basalt and E-glass (B/E/G15) hybrid composite with 15% graphite filler material showed the highest values of 43.81 MPa flexural strength and 3.57 GPa flexural modulus (Fig. 7.7). Many researchers found that the incorporation of filler materials into the composite

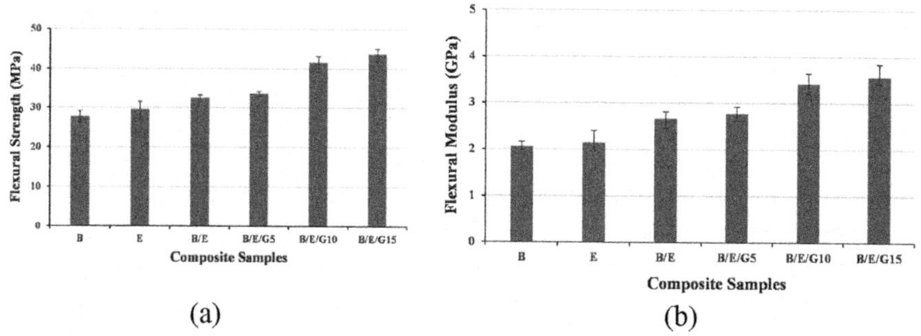

Fig. 7.6 Flexural strength (**a**) and flexural modulus (**b**) of Basalt and E-glass composite samples

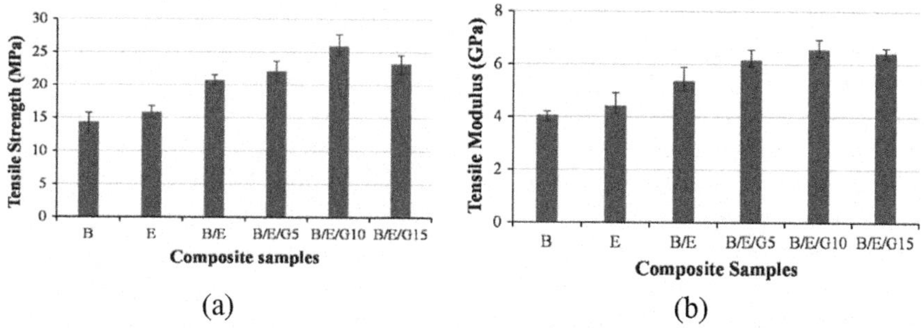

Fig. 7.7 Tensile strength (**a**) and tensile modulus (**b**) of Basalt and E-glass composite samples

material increased both flexural strength and flexural modulus [40, 41]. The reason is attributed to the uniform distribution of filler material and the strong adhesion properties between polymer matrix and filler material. Hence the composites with graphite filler material exhibited superior flexural strength and modulus properties.

Experimental investigations showed that incorporation of the filler material plays a crucial role in improving the mechanical performance of designed novel composite materials.

Tensile properties. The experiments of Basalt and E-glass woven fabric composites were done performing the tensile testing (Fig. 7.7). Thus, the results obtained showed that the impact of filler material in the hybrid composites was quite noticeable and effective. Unlike flexural properties, the tensile strength and tensile modules increased up to 10% with a graphite filler material. Results for Basalt and E-glass (B/E) hybrid composites and Basalt and E-glass hybrid composites with 5% graphite particles (B/E/G5) were almost identical. The epoxy-based Basalt fiber composite showed the least tensile strength and modulus of 14.68 MPa and 4.07 GPa, respectively (Fig. 7.7a, b). Thus, epoxy-based Basalt and E-glass hybrid composites with 10% filler material showed the highest results, i.e. tensile strength of 27.68 MPa (Fig. 7.7a) and tensile modulus of 6.25 GPa (Fig. 7.7b). Hence the composition in B/E/G10 could act as a very good reinforcing filler in the hybrid composites. Also, the internal stresses developed during the test were distributed equally among the fillers and reinforcements.

The presence of a filler material in the composites may improve the adhesion properties and also reduce the porosity content. The hybrid composites in the presence of graphite filler material possessed great strength and modules over the individual fiber composites. Hence, these exceptional properties lead to a potential replacement of other conventional materials used for the microfluidic, in the formation of various dimension groves and channels or in absorbent-coated microchannels.

Impact properties. Further, investigations were done on the impact energy, absorbed by the composite samples (Fig. 7.8). It was defined that the amount of energy was absorbed by an object, when an external load (impact load or sudden load) was applied to it. Tests were conducted using a Charpy impact tester at room temperature with a relative humidity of 30%. The results followed previously obtained results, i.e. the impact energy increased with an increase of filler material. The pure Basalt composites showed the least impact energy at 3.1 kJ/m^2 and B/E/G15 hybrid composites showed the highest at 4.9 kJ/m^2. Furthermore, the B/E hybrid composites and B/E/G5 hybrid composites had almost the same impact on energy. Composite B/E/G15 exhibited the highest impact energy. And the pure composites showed the lowest impact energy values (Fig. 7.9). This rise in hybrid composites is the result of energy-dissipating fiber-related phenomena like fiber debonding, pull out, bridging, and fracture, which cause the resin matrix to flex plastically prior to failure. A set of glass fibers with a length greater than the crucial value for effective reinforcement is likely to result in bridging and fiber fractures, while a set of glass fibers with a length less than the critical value is likely to result in debonding and fiber pull-out.

Moisture analysis. To measure the amount of moisture absorbed by the composites, a moisture analysis test was conducted (Fig. 7.9). After fabrication, the composite

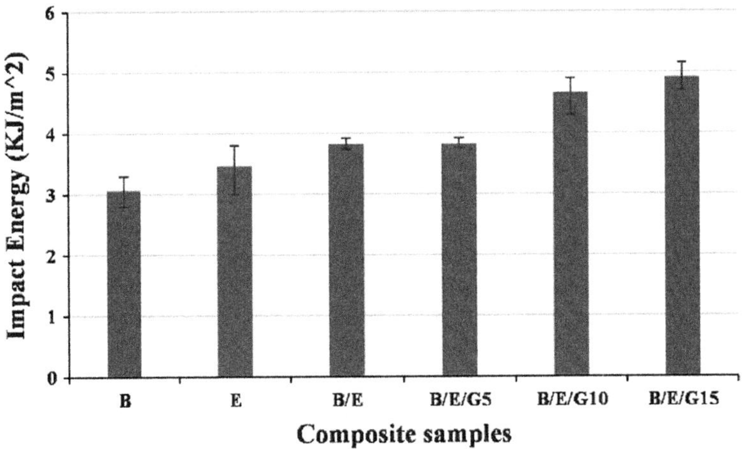

Fig. 7.8 Impact energy of Basalt and E-glass composite samples

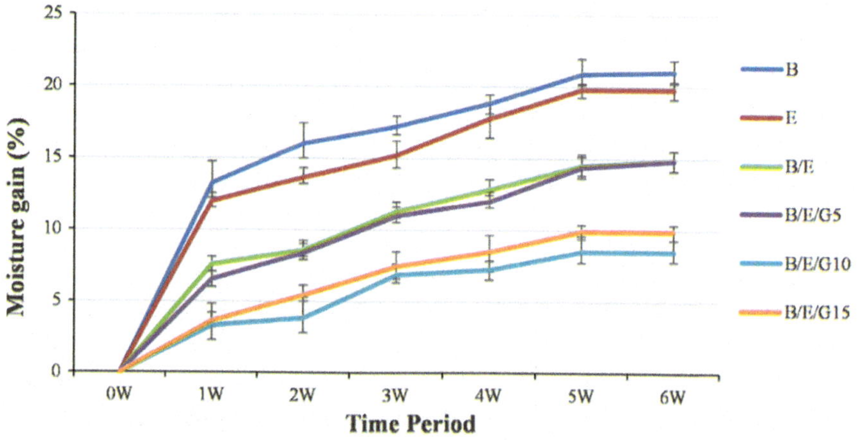

Fig. 7.9 Moisture absorption of Basalt and E-glass hybrid composites

samples were placed in an electric oven to eliminate the moisture in them. To analyze the moisture properties, the composite samples were placed in a jar containing distilled water. The readings were noted by differentiating the initial weight and the final weight (after moisture gain) of the composite samples for six weeks. After six weeks, results showed that the presence of the filler material in a composite sample absorbed less moisture (Fig. 7.9), because the filler material occupied the porosity content in a composite. Thus, pure samples absorbed more moisture than hybrid composites, meaning that the hybrid composites with filler particles have the property of repelling the water, when it gets in contact with it. This occurrence was due to unbalanced molecular forces in the water and the solid interface brings out surface tension. It was observed, that the increase of moisture gained with time in all the composites, and after about five weeks it reached a steady state.

Morphological studies. The surface morphology and porosity formation in Basalt and E-glass woven fiber hybrid composites were studied using the SEM. Thus, the presence of pores in a composite may affect the mechanical performance of a composite material. Previous researchers focused on evacuating the air trapped in voids, by keeping the volatiles in solution, and resin intrusion of evacuated pore spaces. But these methods can drive additional problems such as shrinkage and are not suitable for all composites [42, 43]. Hence, the graphite powder particles were introduced into the composite material to avoid the microbial effect in it. SEM results showed, that the presence of pores in the form of spherical, cylindrical, and helical shapes were observed in the hybrid composites without filler particles (Fig. 7.11a–c). The percentage of porosity in hybrid composites with graphite filler material was smaller when compared to the pure composites (Fig. 10d–f). Most of the pores were replaced with nanoparticles in the B/E/G hybrid composites and that can be attributed to improve the mechanical performance and adhesion properties.

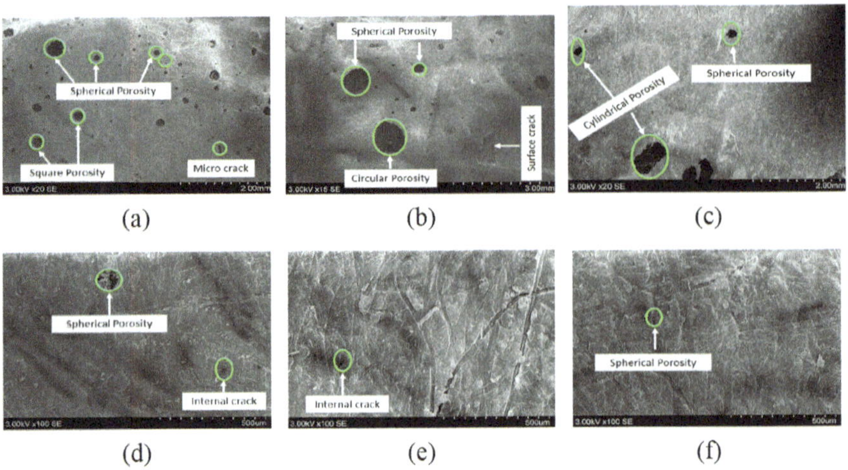

Fig. 7.10 Porosity content in Basalt and E-glass woven fiber hybrid composites **a** B, **b** E, **c** B/E, **d** B/E/G5, **e** B/E/G10, and **f** B/E/G15

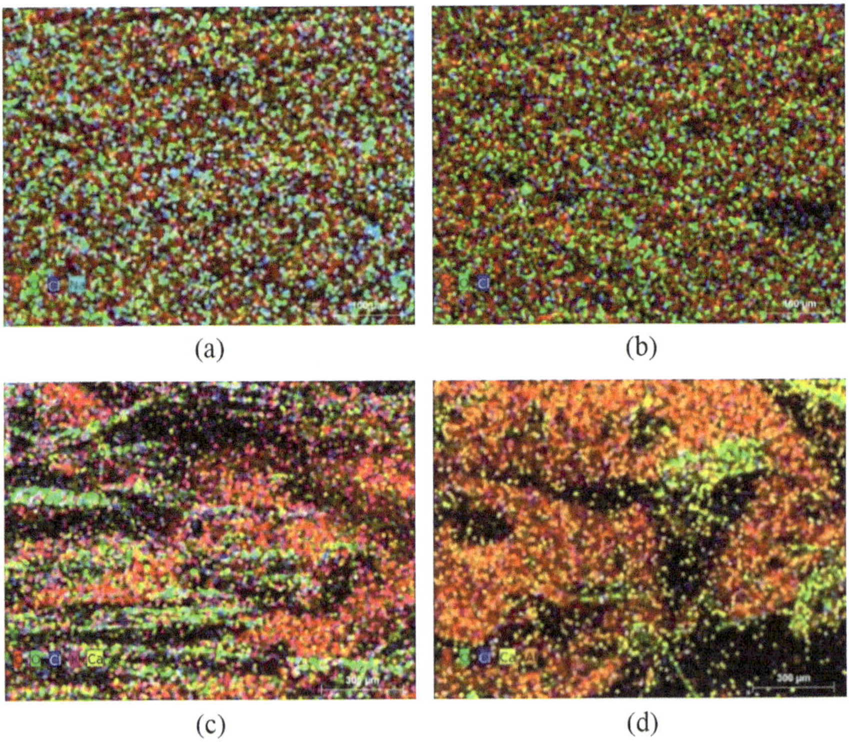

Fig. 7.11 SEM images for map data of hybrid composites **a** B/E, at surface, **b** B/E/G10 at surface, **c** B/E at fracture, **d** B/E/G10 at fracture

The morphological studies were conducted on fractured hybrid samples to understand the chemical composition of the composites. It was observed from SEM–EDX results that hybrid composites exhibited good adhesion properties as well as the surface roughness on composite materials (Fig. 7.11). The map data of chemical composition allowed to evaluate the dominant materials in designed composites: the color notations of red for carbon (C), green for oxygen (O), blue for chlorine (Ci), yellow for calcium (Ca), light blue for sodium (Na), and orange for aluminum (Al), respectively. The dominance of carbon in the B/E/G10 hybrid composite was observed clearly, i.e. it was more at fracture over the surface position. In the B/E hybrid case, the presence of oxygen on the surface position was observed. Thus, a detailed description of atomic and weight concentrations was given in the Table 7.3.

The chemical composition of Basalt and E-glass woven fabric hybrid composites with and without filler materials was analyzed by using SEM–EDX spectroscopy, too. The chemical composition of a hybrid composite at surface and fracture position was observed for comparison purposes. The highest peak was observed for carbon followed by oxygen. Results stated that carbon percentage was higher at a fractured place than surface position (Fig. 7.12).

Due to the presence of graphite particles in B/E/10G, the percentage of carbon reached 78.09% at the fracture position, whereas at the surface position 67.72% (Table 7.3).

Overall, results proved that the mechanical performance of Basalt and E-glass woven fabric hybrid composites increased with the addition of filler material. The effect of the hybridization concept on woven fiber laminates was quite impressive. Thus, designed novel hybrid composites in the presence of filler material can be useful for various engineering applications such as automobiles, construction, biomedical,

Table 7.3 The chemical composition of Basalt and E-glass hybrid composites at surface and fracture position

Element	B/E surface		B/E fracture		B/E/G10 surface		B/E/G10 fracture	
	Weight (%)	Atom. (%)	Weight (%)	Atom. (%)	Weight (%)	Atom. (%)	Weight (%)	Atom. (%)
Carbon (C)	39.85	48.31	53.56	58.98	67.62	73.24	78.09	81.86
Oxygen (O)	58.77	51.10	45.18	40.57	31.89	26.51	10.85	16.74
Calcium (Ca)	0.00	0.00	0.42	0.14	0.24	0.11	0.08	0.10
Chlorine (Cl)	0.94	0.39	0.00	0.00	0.17	0.08	1.92	1.28
Sodium (Na)	0.44	0.20	0.28	0.13	0.08	0.06	0.00	0.00
Silicon (Si)	0.00	0.00	0.56	0.18	0.00	0.00	0.06	0.02

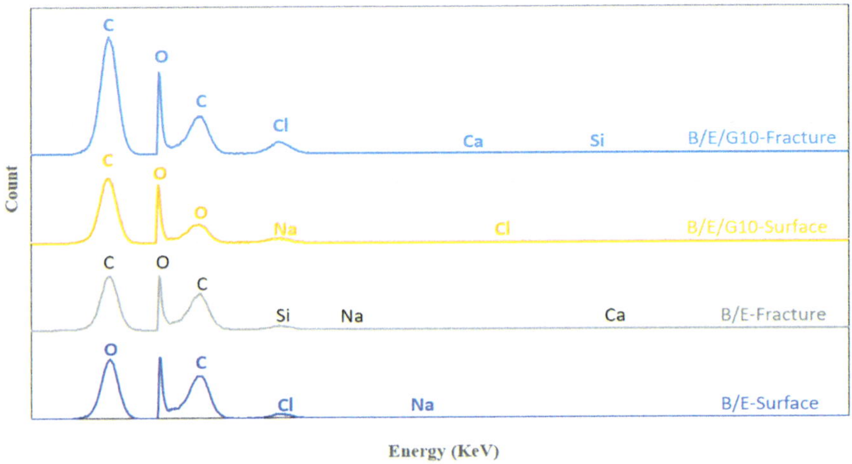

Fig. 7.12 SEM–EDX spectrum analysis on hybrid composites at surface and fracture position

and microfluidic. And deep research of microchannels formation was left for the future research.

7.4 Conclusions

Basalt and E-glass woven fiber-based epoxy hybrid composites were fabricated with graphite filler material utilizing a vacuum-assisted free lamination compression molding fabrication technique. These novel hybrid composites demonstrated enhanced mechanical properties compared to individual Basalt and E-glass fiber composites. The incorporation of filler material into the composite material resulted in improved mechanical performance. It was observed that all composites exhibited hydrophilic surface properties with a contact angle of <90°, with the contact angle increasing as the filler material content in the composite increased. Based on the flexural results, the individual Basalt fiber composites presented the lowest flexural strength of 27.82 MPa and a flexural modulus of 2.06 GPa. In contrast, the B/E/G15 composite exhibited the highest flexural strength at 43.81 MPa and a flexural modulus of 3.57 GPa. The tensile results indicated that the epoxy-based B/E/10G composites exhibited the highest tensile properties, while the epoxy-based Basalt composites showed lower tensile properties. The impact energy observed for individual Basalt and E-glass was 3.1 and 3.5 kJ/m^2, respectively. The moisture uptake of the composites was examined using moisture analysis, and it was noted that all composites followed the same trend up to 5 weeks and remained consistent even after a 6-week period. Furthermore, the porosity content in hybrid composites with filler material was lower than in composites without filler material. The presence of filler material in hybrid composites resulted in a marked increase in carbon weight concentration, demonstrating that the inclusion of filler material and the application of the hybridization concept resulted in superior mechanical performance.

References

1. Atmakuri A, Kolli L, Palevicius A, Urbaite S, Janusas G (2022) Influence of filler materials on wettability and mechanical properties of Basalt/E-glass woven fabric–reinforced composites for microfluidics. Micromach Basel MDPI 13(11):1–14. https://doi.org/10.3390/mi13111875
2. Barua A, Jeet S, Bagal DK, Satapathy P, Agrawal PK (2019) Evaluation of mechanical behavior of hybrid natural fiber reinforced nano sic particles composite using hybrid taguchi-cocoso method. Int J Innov Technol Explor Eng 8:3341–3345
3. Bharath KN, Basavarajappa S (2016) Applications of biocomposite materials based on natural fibers from renewable resources: a review. Sci Eng Compos Mater 23:123–133
4. Summerscales J, Virk A, Hall W (2013) A review of bast fibres and their composites: part 3—modelling. Compos A Appl Sci Manuf 44:132–139
5. Sanjay MR, Arpitha GR, Naik LL, Gopalakrishna K, Yogesha B (2016) Applications of natural fibers and its composites: an overview. Nat Resour 07:108–114
6. Nisini E, Santulli C, Liverani A (2017) Mechanical and impact characterization of hybrid composite laminates with carbon, basalt and flax fibres. Compos B Eng 127:92–99
7. Moudood A, Rahman A, Huq NML, Ochsner A, Islam MM, Francucci G (2020) Mechanical properties of flax fiber-reinforced composites at different relative humidities: experimental, geometric, and displacement potential function approaches. Polym compos 41:4963–4973
8. Nurazzi NM, Asyraf MRR, Fatimah-Athiyah S, Shazleen SS, Ayu-Rafiqah S, Harussani MM, Kamarudin SH, Razman MR, Rahmah M, Zainudin ES, Iiyas RA, Aisyah HA, Norrahim MNF, Abdullah N, Sapuan SM, Khalina A (2021) A review on mechanical performance of hybrid natural fiber polymer composites for structural applications. Polymers 13:2170
9. Rana A, Mandal A, Bandyopadhyay S (2003) Short jute fiber reinforced polypropylene composites: effect of compatibiliser, impact modifier and fiber loading. Compos Sci Technol 63:801–806
10. Jamshaid H, Mishra R, Basra S, Rajput AW, Hassan T, Petru M (2022) Lignocellulosic natural fiber reinforced bisphenol F epoxy based bio-composites: characterization of mechanical electrical performance. J Nat Fibers 19:3317–3332
11. Ramamoorthy SK, Skrifvars M, Persson A (2015) A review of natural fibers used in biocomposites: plant, animal and regenerated cellulose fibers. Polym Rev 55:107–162
12. Ramadan N, Taha I, Hammouda R, Abdellatif MH (2017) Behaviour of hybrid SiC/jute epoxy composites manufactured by vacuum assisted resin infusion. Polym Polym Compos 25:333–344
13. Tang J, Swolfs Y, Longana ML, Yu H, Wisnom MR, Lomov SV, Gorbatikh L (2019) Hybrid composites of aligned discontinuous carbon fibers and self-reinforced polypropylene under tensile loading. Compos A Appl Sci Manuf 123:97–107
14. Aji IS, Zainudin E, Abdan K, Sapuan SM, Khairul MD (2013) Mechanical properties and water absorption behavior of hybridized kenaf/pineapple leaf fibre-reinforced high-density polyethylene composite. J Compos Mater 47:979–990
15. Patel Y, Janusas G, Palevicius A, Vilkauskas A (2020) Development of nanoporous AAO membrane for nano filtration using the acoustophoresis method. Sensors 20(14):3833
16. Yoon JY (2008) Open-surface digital microfluidics. Open Biotechnol J 2:94–100
17. Abdelgawad M, Park P, Wheeler AR (2009) Optimization of device geometry in single-plate digital microfluidics. J Appl Phys 105:1–9
18. Zida SI, Lin Y, Khung YL (2021) Current trends on surface acoustic wave biosensors. Adv Mater Technol 6(6):1–12

19. Guo J, Lin L, Zhao K, Song Y, Huang M, Zhu Z, Zhou L, Yang C (2020) Auto-affitech: an automated ligand binding affinity evaluation platform using digital microfluidics with a bidirectional magnetic separation method. Lab Chip 20:1577–1585
20. Chen W, Hao H, Jong M, Cui J, Shi Y, Chen L, Pham TM (2017) Quasi-static and dynamic tensile properties of basalt fibre reinforced polymer. Compos B Eng 125:123–133
21. Lopresto V, Leone C, De Iorio I (2011) Mechanical characterisation of basalt fibre reinforced plastic. Compos B Eng 42(4):717–723
22. Seghini MC, Touchard F, Sarasini F, Cech V, Chocinski-Arnault L, Mellier D, Tirllo J, Bracciale MP, Zvonek M (2019) Engineering the interfacial adhesion in basalt/epoxy composites by plasma polymerization. Compos A Appl Sci Manuf 122:67–76
23. Kuranska M, Barczewski M, Uram K, Lewandowski K, Prociak A, Michatowski S (2019) Basalt waste management in the production of highly effective porous polyurethane composites for thermal insulating applications. Polym Test 76:90–100
24. Matykiewicz D, Barczewski M, Michalowski S (2019) Basalt powder as an eco-friendly filler for epoxy composites: thermal and thermo-mechanical properties assessment. Compos B Eng 164:272–279
25. Dorigato A, Pegoretti A (2014) Flexural and impact behaviour of carbon/basalt fibers hybrid laminates. J Compos Mater 48:1121–1130
26. Sakthivel M, Vijayakumar S, Prasad NK (2015) Drilling analysis on basalt/sisal reinforced polymer composites using ANOVA and regression model. Appl Math Sci 9:3285–3290
27. Sabet SMM, Akhlaghi F, Eslami-Farsani R (2015) The effect of thermal treatment on tensile properties of basalt fibers. J Ceram Sci Technol. 6(3):245–248
28. Sarasini F, Tirillo J, Valente M, Valente T, Cioffi S, Iannace S, Sorrentino L (2013) Effect of basalt fiber hybridization on the impact behavior under low impact velocity of glass/basalt woven fabric/epoxy resin composites. Compos A Appl Sci Manuf 47:109–123
29. Amaro AM, Reis PNB, de Moura MFSF (2006) Residual strength after low velocity impact in carbon-epoxy laminates. Mater Sci Forum 514:624–628
30. Chenghong H, Yubin L, Zuoguang Z, Zhijie S (2008) Impact damage modes and residual flexural properties of composites beam. J Reinf Plast Compos 27:1163–1175
31. Santiuste C, Sanchez-Saez S, Barbero E (2010) Residual flexural strength after low-velocity impact in glass/polyester composite beams. Compos Struct 92(1):25–30
32. Shim VPW, Yang LM (2005) Characterization of the residual mechanical properties of woven fabric reinforced composites after low-velocity impact. Int J Mech Sci 47:647–665
33. Valentino P, Sgambitterra E, Furgiuele F, Romano M, Ehrlich I, Gebbeken N (2014) Mechanical characterization of basalt woven fabric composites: numerical and experimental investigation. Frat Ed Integrita Strutt 28(8):1–10
34. King FL, Kumar AAJ, Vijayaragahavan S (2019) Mechanical characterization of polylactic acid reinforced bagasse/basalt hybrid fiber composites. J Compos Mater 53(1):33–43
35. Harsha-Vardhan VSS, Babu KRP, Raghu-Kumar B, Rama-Krishna G (2020) Mechanical characterization of E-glass fibre reinforced composite with graphene nano particles as filler. Int J Emerg Trends Eng Res 8(9):5043–5047
36. Arun Prakash VR, Rajadurai A (2016) Thermo-mechanical characterization of siliconized E-glass fiber/hematite particles reinforced epoxy resin hybrid composite. Appl Surf Sci 384:99–106
37. Caylak S, Ertas M, Donmez Cavdar A, Angin N (2021) Mechanical characteristics and hydrophobicity of alkyl ketene dimer compatibilized hybrid biopolymer composites. Polym Compos 42:2324–2333
38. Meng F, Zhang Y, Xiong Z, Wang G, Li F, Zhang L (2018) Mechanical, hydrophobic and thermal properties of an organic-inorganic hybrid carrageenan-polyvinyl alcohol composite film. Compos B Eng 143:1–8
39. Wang F, Wu Y, Huang Y (2018) Novel application of graphene oxide to improve hydrophilicity and mechanical strength of aramid nanofiber hybrid membrane. Compos A Appl Sci Manuf 110:126–132

40. Tarawneh MA, Ahmad SH, Yahya SY, Rasid R, Noum SYE (2011) Mechanical properties of thermoplastic natural rubber reinforced with multi-walled carbon nanotubes. J Reinforced Plast Compos 30(4):363–368
41. Sreekanth MS, Joseph S, Mhaske ST, Mahanwar PA (2011) Effects of mica and fly ash concentration on the properties of polyester thermoplastic elastomer composites. J Thermoplast Compos Mater 24:317–330
42. Malysheva GV, Guzeva TA (2021) Technological support for porosity reduction of polymer composite material components. Technol Met 13:1692–1695
43. Pollard M, Tran P, Dickens T (2020) Porosity reducing processing stages of additive manufactured molding (Amm) for closed-mold composite fabrication. Materials 13(23):5328

Chapter 8
Practical Application of Natural Fiber-Based Hybrid Composite

Abstract Due to global climate issues, many efforts have been made to provide biodegradable and environmentally friendly materials for the next era of composite materials. Due to the increasing awareness of natural fiber composites, they are being used in various applications to limit the emission of ozone-depleting substances. Green composites are great alternatives to petroleum-based materials that are destructive to the planet. In addition to many advantages of natural fiber composites, they present several difficulties in use, such as high moisture content and poor similarity between fibers and mesh, poor fire resistance, lower durability and strength compared to weaker natural fibers. This chapter provides a recent study on the types of natural fibers and their applications, so that green composites can be used in various applications to better understand.

8.1 Introduction

Fiber-reinforced composites are better than pure polymer composites. There should be an occurrence of auxiliary applications on account of its high mechanical properties and lower expense [1]. Due to an account of advanced properties, hybrid composites are a replacement for synthetic composites, particularly when it comes to the load bearing and construction applications. A recent study stated that goods made up of natural fiber composites have been increased and used in several applications [2]. The utilization of natural fiber composites in place of synthetic and petroleum-based fiber composites has increased drastically in various sectors such as sports equipment, aerospace, office machinery, automobile, and shipping industries are one of the main application sectors for lightweight composite materials [3]. The researchers found interest in the replacement of bio fiber-based composites due to their abundance, cost-effectiveness, recyclability, eco friendliness, low fabrication cost and biodegradability [4].

A recent survey stated that the fabrication of fiber-reinforced polymer composites has become more popular in commercial applications, especially in the aerospace industry. The utilization of fiber-based composite caused a reduction in overall weight

A. Atmakuri et al., *Fiber-Based Hybrid Composites for Engineering*, Springer Series on Polymer and Composite Materials, https://doi.org/10.1007/978-3-031-93144-4_8

by 35% of the aircraft and shows a positive effect on fuel consumption and efficiency of the overall performance [5]. Due to these properties, aircraft industries such as Boeing, Dreamliner and Airbus companies have started using composites in the majority of their components [6]. The main crucial factor in composite fabrication is weight reduction. Some of the natural fibers serve as a natural filler material in the composite fabrication in the presence of bio-based and biodegradable matrix material designed for the extrusion film blowing of thin packages [7].

Kumar et al. [8] worked on natural fiber composites in the replacement of synthetic fiber composites for military applications. They compared natural fiber (pineapple) epoxy composite material and synthetic (aramid) fiber composite material in a hard armor. Results showed that the armor plate made from natural fibers has a stronger impact strength than synthetic fiber plates. Natural fiber hybrid nanocomposites are used in various industries because of their improvement in mechanical, thermal and moisture absorption along with light weight and less cost [9]. Nowadays, these composites are in construction, biomedical, filtration, drug delivery, tissue template, wound dressing and cosmetics [10].

8.2 Applications of Natural Fiber Composites

Hybrid Composites reinforced with natural fibers are quickly becoming a sustainable alternative to materials based on metal or ceramic in various applications, including the construction, biomedical, electronic, automotive, aerospace, and marine industries. Composites made of natural fibers have good specific qualities, but their properties vary greatly. More sophisticated processing of natural fibers and their hybrid composites can and will overcome their frailty. In the "green" materials environment of the twenty-first century, their individual qualities ought to serve as a strong foundation for developing new uses and prospects for biocomposites or natural fiber composites [11]. Utilizing natural fiber composites for a variety of purposes has created new opportunities for industrial sectors and scholars to produce sustainable modules for use in the future. Some of the potential applications are listed below [12].

1. Filter media: Liquid, gas and molecule filtration.
2. Nano-sensors: Biochemical, thermal and piezoelectric sensor.
3. Cosmetics: Skin cleaning and healing.
4. Life science: Drug delivery carrier and wound dressing.
5. Tissue engineering scaffolding: Porous membrane for skin, 3D scaffolds for bone and cartilage regeneration.
6. Industrial applications: Micro/nano electronic devices, electrostatic dissipation, LCD devices, lightweight space craft materials.

The use of natural fiber as reinforcement in the polymer matrix brought environmental awareness to the attention of people worldwide. Combining two or more distinct fiber types in which one type of fiber compensates for the shortcomings of another is known as a hybrid composite. In numerous applications, natural fiber reinforced polymer composites have been shown to be a viable substitute for synthetic fiber reinforced polymer composites [13–15]. Few natural fiber composites have been created, and most of their technologies are still in the research and development stages. In contrast, many natural fiber composite goods are being produced and commercialized. Package shelves, door panels, instrument panels, armrests, headrests, and seat shells are among the automotive components made of natural fiber composites [16].

Wood fiber composites are being used in many applications in decks, docks, window frames and molded panel components. The passenger car bumper beam is manufactured by kenaf/glass epoxy composite material [17, 18]. Recently, banana fiber reinforced composites are coming into interest due to the innovative application of banana fiber in under-floor protection for passenger cars. Automobile parts such as rear-view mirror, visor in two-wheeler, billion seat cover, indicator cover, cover L-side, name plate was fabricated using sisal and roselle fibers hybrid composites [19, 20].

8.3 Conclusions

Strict environmental regulations and the search for sustainable materials have led researchers to develop environmentally friendly materials. Natural fiber hybrid composites with exceptional properties have potential applications in modern industry in general and in the automotive industry and civil infrastructure sector in particular. The comparable mechanical properties of natural fibers to those of synthetic fibers, the low production cost, good thermal and acoustic characteristics and environmentally friendly handling make them a good alternative as reinforcement in polymer composites. Due to their outstanding properties, Natural fiber hybrid composites have good potential in load-bearing applications. The possibilities seem great, as many natural fibers and their polymer blends are still unexplored. In addition, the different native species of many natural fibers also need to be explored. Hybrid polymer composites and the synergistic effect of different natural fibers is an area where future studies can focus.

References

1. Nirmal U, Hashim J, Megat Ahmad MMH (2015) A review on tribological performance of natural fibre polymeric composites. Tribol Int 83:77–104. https://doi.org/10.1016/j.triboint.2014.11.003
2. Singh MK, Tewari R, Zafar S et al (2023) A comprehensive review of various factors for application feasibility of natural fiber-reinforced polymer composites. Results Mater 17:100355. https://doi.org/10.1016/j.rinma.2022.100355

3. Shahinur S, Sayeed MMA, Hasan M et al (2022) Current development and future perspective on natural jute fibers and their biocomposites. Polymers 14:1445. https://doi.org/10.3390/polym14071445
4. Ahmad F, Choi HS, Park MK (2015) A review: natural fiber composites selection in view of mechanical, light weight, and economic properties. Macromol Mater Eng 300:10–24. https://doi.org/10.1002/mame.201400089
5. Saravana-Bavan D, Mohan-Kumar G (2010) Potential use of natural fiber composite materials in India. J Reinf Plast Compos 29:3600–3613. https://doi.org/10.1177/0731684410381151
6. Jaafar J, Siregar JP, Salleh SM et al (2019) Important considerations in manufacturing of natural fiber composites: a review. Int J Precis Eng Manuf Technol 6:647–664. https://doi.org/10.1007/s40684-019-00097-2
7. Pandey JK, Ahn SH, Lee CS et al (2010) Recent advances in the application of natural fiber based composites. Macromol Mater Eng 295:975–989. https://doi.org/10.1002/mame.201000095
8. Kumar B, Agumba DO, Pham DH et al (2021) Recent research progress on lignin-derived resins for natural fiber composite applications. Polymers 13:1162. https://doi.org/10.3390/polym13071162
9. José da Silva L, Hallak-Panzera T, Luis-Christoforo A et al (2012) Numerical and experimental analyses of biocomposites reinforced with natural fibres. Int J Mater Eng 2:43–49. https://doi.org/10.5923/j.ijme.20120204.03
10. Sanjay MR, Arpitha GR, Naik LL et al (2016) Applications of natural fibers and its composites: an overview. Nat Resour 07:108–114. https://doi.org/10.4236/nr.2016.73011
11. Aliotta L, Gigante V, Coltelli MB et al (2019) Evaluation of mechanical and interfacial properties of bio-composites based on poly(lactic acid) with natural cellulose fibers. Int J Mol Sci 20:960. https://doi.org/10.3390/ijms20040960
12. Mohammed L, Ansari MNM, Pua G et al (2015) A review on natural fiber reinforced polymer composite and its applications. Int J Polym Sci 2015:1–15. https://doi.org/10.1155/2015/243947
13. Dong C (2018) Review of natural fibre-reinforced hybrid composites. J Reinf Plast Compos 37:331–348. https://doi.org/10.1177/0731684417745368
14. Sanjay MR, Madhu P, Jawaid M et al (2018) Characterization and properties of natural fiber polymer composites: a comprehensive review. J Clean Prod 172:566–581. https://doi.org/10.1016/j.jclepro.2017.10.101
15. Barua A, Jeet S, Bagal DK et al (2019) Evaluation of mechanical behavior of hybrid natural fiber reinforced nano sic particles composite using hybrid taguchi-cocoso method. Int J Innov Technol Explor Eng 8:3341–3345
16. Luz F, Garcia Filho F, Oliveira M et al (2020) Composites with natural fibers and conventional materials applied in a hard armor: a comparison. Polymers 12:1920. https://doi.org/10.3390/polym12091920
17. Furtado SCR, Araújo AL, Silva A et al (2014) Natural fibre-reinforced composite parts for automotive applications. Int J Automot Compos 1:18. https://doi.org/10.1504/IJAUTOC.2014.064112
18. Abdul Khalil HPS, Davoudpour Y, Islam MN et al (2014) Production and modification of nanofibrillated cellulose using various mechanical processes: a review. Carbohydr Polym 99:649–665. https://doi.org/10.1016/j.carbpol.2013.08.069
19. Adesina OT, Jamiru T, Sadiku ER et al (2019) Mechanical evaluation of hybrid natural fibre–reinforced polymeric composites for automotive bumper beam: a review. Int J Adv Manuf Technol 103:1781–1797. https://doi.org/10.1007/s00170-019-03638-w
20. Suriani MJ, Ilyas RA, Zuhri MYM et al (2021) Critical review of natural fiber reinforced hybrid composites: processing, properties, applications and cost. Polymers 13:3514. https://doi.org/10.3390/polym13203514

The manufacturer's authorised representative in the EU is Springer Nature Customer Service Centre GmbH, Europaplatz 3, 69115 Heidelberg, Germany. If you have any concerns regarding our products, please contact ProductSafety@springernature.com

Printed and bound by CPI Group (UK) Ltd, Croydon, CR0 4YY
30/09/2026
02225604-0008